SpringerBriefs in Mathematics

SpringerBriefs in Mathematics showcases expositions in all areas of mathematics and applied mathematics. Manuscripts presenting new results or a single new result in a classical field, new field, or an emerging topic, applications, or bridges between new results and already published works, are encouraged. The series is intended for mathematicians and applied mathematicians.

More information about this series at http://www.springer.com/series/10030

Toka Diagana • François Ramaroson

Non-Archimedean Operator Theory

 Springer

Toka Diagana
Department of Mathematics
Howard University
Washington, DC, USA

François Ramaroson
Department of Mathematics
Howard University
Washington, DC, USA

ISSN 2191-8198   ISSN 2191-8201   (electronic)
SpringerBriefs in Mathematics
ISBN 978-3-319-27322-8   ISBN 978-3-319-27323-5   (eBook)
DOI 10.1007/978-3-319-27323-5

Library of Congress Control Number: 2015957041

Mathematics Subject Classification (2010): 12J25, 26E30, 32P05, 37P20, 46S10, 47S10, 47A53, 35P05

Springer Cham Heidelberg New York Dordrecht London

Printed on acid-free paper

Springer International Publishing AG Switzerland is part of Springer Science+Business Media (www.
springer.com)

This book is dedicated, by Toka Diagana, to his wife Sally, for her unyielding love, support, and exceptional patience.

This book is dedicated, by François Ramaroson, to his wife Luna, for her unyielding love, support, and exceptional patience.

Preface

This book is focused on the theory of linear operators on non-Archimedean Banach spaces. It is to some extent a sequel of the authors' recent work on linear operators on non-Archimedean Banach spaces as well as their spectral theory. There are not many books exclusively dealing with operator theory on non-Archimedean Banach spaces or their variants, and the authors wish to add to the scarce literature on the subject. A minimum necessary background material has been gathered which will allow a relatively friendly access to the book.

Beginning graduate students who wish to enter the field of non-Archimedean functional analysis should benefit from the material covered, but an expert reader might also find some of the results interesting enough to be sources of inspiration. Prerequisites for the book are the basic courses in classical real and complex analysis and some knowledge of basic functional analysis. Further, knowledge of basic algebra (groups, rings, fields, vector spaces) and some familiarity with p-adic numbers, such as Gouvêa's introductory book *p-adic numbers: An Introduction* (second edition, Springer, 2003), will be a huge plus. The student would gain a long way knowing the first four chapters of the book *Local Fields* by Cassels (1986, London Mathematical Society, Student Texts 3).

After reading this book, the reader might benefit a great deal if he/she moves on forward with deeper material like the book *Spectral Theory and Analytic Geometry over non-Archimedean Fields* by Berkovich which is geared toward a systematic treatment of the spectral theory for Banach•algebras (respectively, Rigid analytic geometry) or the books *Non-Archimedean Functional Analysis* by Schneider and *Non-Archimedean Functional Analysis* by van Rooij. Those readers with more arithmetical inclinations will profit from the book *p-adic Analysis: A Short Course on Recent Work* by Koblitz.

The topics treated in the book range from a basic introduction to non-Archimedean valued fields, free Banach spaces, and (possibly unbounded) linear operators in the non-Archimedean setting to the spectral theory for some classes of linear operators and their perturbations. Although some parts of the material are taken from the book *Non-Archimedean Linear Operators and Applications* by Diagana, this book is more comprehensive as it covers many new topics. It emphasizes the role of the theory of Fredholm operators which is used as an important tool. This approach in the study of the spectral theory of linear operators should play more roles in larger context than the ones covered in the book, and in this regard, the book is a good introduction to the spectral theory of linear operators in the non-Archimedean setting. Explicit descriptions of the spectra of some linear operators are worked out.

Chapter 1 is of a background nature. It covers non-Archimedean valued fields and contains many details and examples on non-Archimedean valuations, the topology induced by these valuations, and their extensions. Spherical completeness is defined and some related properties are proved and illustrated with examples. The Krull valuation is introduced.

Chapter 2 is also of a background nature and covers non-Archimedean Banach spaces. These spaces are complete normed vector spaces over a complete non-Archimedean valued field. Of special interest are the free Banach spaces, especially the p-adic Hilbert space, and they are studied in detail. A structure theorem for the p-adic Hilbert space is proved.

Chapter 3 is on the bounded linear operators. Various properties are stated and proved. Finite rank operators, completely continuous operators, and Fredholm operators are all discussed with a view toward the applications in spectral theory.

Chapter 4 introduces and studies properties of the Shnirel'man integral. Among other things, such an integral is used to construct the so-called Vishik spectral theorem.

Chapter 5 contains the determination of the spectrum of a perturbation of a bounded diagonal operator by finite rank operators. The technique uses the theory of Fredholm operators.

Chapter 6 treats general unbounded operators, closed operators, and the spectrum of unbounded operators and the unbounded Fredholm operators.

Chapter 7 is devoted to the study of spectral theory for the perturbations of an unbounded operator by operators of finite rank or by completely continuous operators. Special emphasis is put on the computation of the essential spectrum of these perturbed unbounded linear operators.

This book is intended for graduate and postgraduate students, mathematicians, and nonmathematicians such as physicists and engineers who are interested in functional analysis in the non-Archimedean context. Further, it can be used as an introduction to the study of linear operators in general and to the study of spectral theory in other special cases.

Washington, D.C, USA   Toka Diagana
May, 2015   François Ramaroson

Acknowledgments

The authors would like to express their thanks to colleagues, students, and collaborators who have helped in the course of this work.

Contents

1 Non-Archimedean Valued Fields 1
  1.1 Valuation 1
    1.1.1 Definitions and First Properties 1
    1.1.2 The Topology Induced by a Valuation on $\mathbb{K}$ 5
    1.1.3 Non-Archimedean Valuations 8
    1.1.4 Some Analysis on a Complete Non-Archimedean
           Valued Field 13
    1.1.5 The Order Function for a Discrete Valuation 15
  1.2 Examples 16
    1.2.1 Examples of Archimedean Valuation 16
    1.2.2 Examples of Non-Archimedean Valued Fields 17
  1.3 Additional Properties of Non-Archimedean
       Valued Fields 28
  1.4 Some Remarks on Krull Valuations 36
  1.5 Bibliographical Notes 39

2 Non-Archimedean Banach Spaces 41
  2.1 Non-Archimedean Norms 41
  2.2 Non-Archimedean Banach Spaces 44
  2.3 Free Banach Spaces 50
  2.4 The p-adic Hilbert Space $\mathbb{E}_\omega$ 54
  2.5 Bibliographical Notes 60

3 Bounded Linear Operators in Non-Archimedean Banach Spaces 61
  3.1 Bounded Linear Operators 61
    3.1.1 Definitions and Examples 61
    3.1.2 Basic Properties 64
    3.1.3 Bounded Linear Operators in Free Banach Spaces 65
  3.2 Additional Properties of Bounded Linear Operators 67
    3.2.1 The Inverse Operator 67

   3.2.2 Perturbations of Orthogonal Bases Using the
      Inverse Operator... 68
   3.2.3 The Adjoint Operator 73
 3.3 Finite Rank Linear Operators 75
   3.3.1 Basic Definitions 75
   3.3.2 Properties of Finite Rank Operators........................... 75
 3.4 Completely Continuous Linear Operators............................ 77
   3.4.1 Basic Properties 77
   3.4.2 Completely Continuous Linear Operators on $\mathbb{E}_\omega$ 77
 3.5 Bounded Fredholm Linear Operators.................................. 78
   3.5.1 Definitions and Examples...................................... 78
   3.5.2 Properties of Fredholm Operators 79
 3.6 Spectral Theory for Bounded Linear Operators...................... 81
   3.6.1 The Spectrum of a Bounded Linear Operator................. 81
   3.6.2 The Essential Spectrum of a Bounded Linear Operator 82
 3.7 Bibliographical Notes 84

4 **The Vishik Spectral Theorem** 85
 4.1 The Shnirel'man Integral and Its Properties 85
   4.1.1 Basic Definitions 85
   4.1.2 The Shnirel'man Integral 91
 4.2 Distributions with Compact Support 96
 4.3 Cauchy–Stieltjes and Vishik Transforms............................ 98
 4.4 Analytic Bounded Linear Operators.................................. 102
 4.5 Vishik Spectral Theorem 104
 4.6 Bibliographical Notes 105

5 **Spectral Theory for Perturbations of Bounded Diagonal**
 Linear Operators... 107
 5.1 Spectral Theory for Finite Rank Perturbations
   of Diagonal Operators.. 107
   5.1.1 Introduction... 107
   5.1.2 Spectral Analysis for the Class of Operators $T = D + K$.... 109
   5.1.3 Spectral Analysis for the Class of Operators $T = D + F$.... 112
 5.2 Computation of $\sigma_e(D)$ 115
 5.3 Spectrum of $T = D + F$ 118
 5.4 Examples... 118
 5.5 Bibliographical Notes 121

6 **Unbounded Linear Operators** 123
 6.1 Unbounded Linear Operators on a Non-archimedean
   Banach Space 123
 6.2 Closed Linear Operators 124
 6.3 The Spectrum of an Unbounded Operator 126
 6.4 Unbounded Fredholm Operators..................................... 127
 6.5 Bibliographical Notes 129

7   Spectral Theory for Perturbations of Unbounded Linear Operators .. 131
   7.1   Introduction 131
   7.2   Spectral Analysis for the Class of Operators $T = D + K$ 132
   7.3   Spectral Analysis for the Class of Operators $T = D + F$ 133
   7.4   Computation of $\sigma_e(D)$ 135
   7.5   Main Result 138
   7.6   Bibliographical Notes 139

A   The Shnirel'man Integral 141
   A.1   Distributions with Compact Support 146
   A.2   Cauchy-Stieltjes and Vishik Transforms 149

References 151

Index 155

Chapter 1
Non-Archimedean Valued Fields

In the classical settings of the field of complex numbers $\mathbb{C}$ and the field of real numbers $\mathbb{R}$, the absolute value plays an important role in the Topology and in the Analysis on objects over these fields. In this chapter, we generalize the absolute value by introducing the notion of *valuation* on a general field $\mathbb{K}$. As we shall see, this notion of valuation allows one to have a natural topology on the field itself and also on objects that are defined over the field. Analysis on the field and on these objects follows naturally. As it turns out, there are two kinds of valuation, one is the *archimedean valuation*, as in the cases of $\mathbb{C}$ and $\mathbb{R}$, and the other is the *non-archimedean valuation*. In this book, our focus will be on the non-archimedean valuation. More specifically, we will work on free Banach spaces over non-archimedean valued fields and Operator theory on them. In this chapter we shall first develop the theory of valuation and then we shall give many examples to illustrate the theory. This chapter will mostly serve as background for the theory of operators that will be developed later in the book. Most of the results are well-known and we gather only those that will serve our purposes.

1.1 Valuation

We mostly follow Artin [2] and Cassels [10] in the presentation of the theory of valuation on a field.

1.1.1 Definitions and First Properties

Definition 1.1. Let $\mathbb{K}$ be a field. A valuation on $\mathbb{K}$ is a map $|\cdot| : \mathbb{K} \to \mathbb{R}$ such that for some real number $C \geq 1$, the following hold:

© The Author(s) 2016
T. Diagana, F. Ramaroson, *Non-Archimedean Operator Theory*,
SpringerBriefs in Mathematics, DOI 10.1007/978-3-319-27323-5_1

(1) $|x| \geq 0$ for any x in $\mathbb{K}$ with equality only for $x = 0$.
(2) $|xy| = |x| \cdot |y|$ for any x, y in $\mathbb{K}$.
(3) For x in $\mathbb{K}$ if $|x| \leq 1$, then $|x + 1| \leq C$.

The valuation $|\cdot|$ such that $|x| = 1$ for every non-zero x and $|0| = 0$ is called the *trivial valuation*.

Proposition 1.2. *The following hold:*

(1) $|1| = 1$.
(2) For x in $\mathbb{K}$, if $|x^n| = 1$ then $|x| = 1$.
(3) $|-1| = 1$.
(4) $|-x| = |x|$.

Proposition 1.3. *Let $|\cdot| : \mathbb{K} \to \mathbb{R}$ be a valuation on $\mathbb{K}$ and λ a positive real number, then $|\cdot|_\lambda$ defined by*

$$\left| x \right|_\lambda := \left| x \right|^\lambda$$

for any x in $\mathbb{K}$ is a valuation on $\mathbb{K}$.

Proof. Properties (1) and (2) of Definition 1.1 are clear. For (3) of Definition 1.1, if $|x|_\lambda \leq 1$ then $|x|^\lambda \leq 1$, hence $|x| \leq 1$, and since $|\cdot|$ is a valuation, $|x + 1| \leq C$ and $|x + 1|_\lambda = |x + 1|^\lambda \leq C^\lambda$ hence (3) of Definition 1.1 holds with the constant C^λ.

Definition 1.4. Two valuations $|\cdot|_1$ and $|\cdot|_2$ on the field $\mathbb{K}$ are equivalent if there exists a positive real numbers λ such that $|\cdot|_2 = |\cdot|_1^\lambda$.

This is an equivalence relation on the set of valuations on the field $\mathbb{K}$.

Definition 1.5. A valuation $|\cdot|$ on the field $\mathbb{K}$ satisfies the *triangle inequality* if for any x, y in $\mathbb{K}$,

$$\left| x + y \right| \leq \left| x \right| + \left| y \right|.$$

Proposition 1.6. *Let $|\cdot|$ be a valuation on $\mathbb{K}$, then, it satisfies the triangle inequality if and only if one can take $C = 2$ in (3) of Definition 1.1.*

Suppose the valuation satisfies the triangle inequality and let x be such that $|x| \leq 1$ then

$$\left| x + 1 \right| \leq \left| x \right| + \left| 1 \right| \leq 2.$$

Conversely, suppose that one can take $C = 2$ in (3) of Definition 1.1. The following lemmas are needed:

Lemma 1.7. *For any positive integer n and for any $x_1, \ldots, x_{2^n}$ in $\mathbb{K}$,*

$$\left| x_1 + x_2 + \ldots + x_{2^n} \right| \leq 2^n \max \left\{ |x_j| : 1 \leq j \leq 2^n \right\}.$$

Proof. We use induction on n. For $n = 1$, if $|x_1| \leq |x_2|$ then $\left| \frac{x_1}{x_2} \right| \leq 1$, hence $\left| \frac{x_1}{x_2} + 1 \right| \leq 2$ and therefore $|x_1 + x_2| \leq 2|x_2| = 2 \max\{|x_1|, |x_2|\}$. The case $|x_2| \leq |x_1|$ is handled similarly. Assume the result true for $n - 1$, then for $x_1, \ldots, x_{2^n}$ in $\mathbb{K}$,

$$\left| x_1 + \ldots + x_{2^n} \right| \leq 2 \max \left\{ |x_1 + \ldots + x_{2^{n-1}}|, |x_{2^{n-1}+1} + \ldots + x_{2^n}| \right\}.$$

Putting $y_1 = x_{2^{n-1}+1}$, $y_2 = x_{2^{n-1}+2}$, $\ldots$, $y_{2^{n-1}} = x_{2^{n-1}+2^{n-1}} = x_{2^n}$ and using the inductive hypothesis on $y_1, y_2, \ldots y_{2^{n-1}}$ and $x_1, \ldots, x_{2^{n-1}}$ yields Lemma 1.7.

Lemma 1.8. *For any positive integer N and for any $x_1, \ldots, x_N$ in $\mathbb{K}$,*

$$\left| x_1 + x_2 + \ldots + x_N \right| \leq 2N \max \left\{ |x_j| : 1 \leq j \leq N \right\}.$$

Proof. Let n be such that $2^{n-1} < N \leq 2^n$ so that $2^n < 2N$. Putting $x_{N+1} = x_{N+2} = \ldots = x_{2^n} = 0$ one obtains,

$$\left| x_1 + x_2 + \ldots + x_N \right| = \left| x_1 + x_2 + \ldots + x_N + x_{N+1} + \ldots + x_{2^n} \right|$$

$$\leq 2^n \max \left\{ |x_j| : 1 \leq j \leq 2^n \right\}$$

$$\leq 2N \max \left\{ |x_j| : 1 \leq j \leq N \right\}.$$

As a consequence of Lemma 1.8, upon putting $x_1 = \ldots = x_N = 1$ one obtains:

Corollary 1.9. *For any positive integer N, $|N| \leq 2N$.*

We can now finish the proof of Proposition 1.6. Let x, y be in $\mathbb{K}$, then for any positive integer n,

$$\left| x + y \right|^n = \left| (x + y)^n \right|$$

$$= \left| \sum_{j=1}^{n} \binom{n}{j} x^j y^{n-j} \right|$$

$$\leq 2(n+1) \max \left\{ \left| \binom{n}{j} \right| |x|^j |y|^{n-j} : 0 \leq j \leq n \right\}$$

$$= 4(n+1) \max \left\{ \binom{n}{j} |x|^j |y|^{n-j} : 0 \leq j \leq n \right\}$$

$$\leq 4(n+1) \sum_{j=0}^{n} \binom{n}{j} \left|x\right|^{j} \left|y\right|^{n-j}$$

$$= 4(n+1)\left(\left|x\right| + \left|y\right|\right)^{n}.$$

Taking n-th root yields

$$\left|x+y\right| \leq \left(4(n+1)\right)^{\frac{1}{n}}\left(\left|x\right| + \left|y\right|\right).$$

Now letting $n \to \infty$ we find

$$\left|x+y\right| \leq \left|x\right| + \left|y\right|$$

as desired.

Proposition 1.10. *Every valuation on* $\mathbb{K}$ *is equivalent to one that satisfies the triangle inequality.*

Proof. Let $|\cdot|$ be a valuation on $\mathbb{K}$ and let $C \geq 1$ be the associated constant. Let $\lambda = log_2 C$, and consider $|\cdot|_\lambda = |\cdot|^{\frac{1}{\lambda}}$, then if $|x|^{\frac{1}{\lambda}} \leq 1$, then $|x| \leq 1$ and $|x+1| \leq C$, hence $|x+1|^{\frac{1}{\lambda}} \leq C^{\frac{1}{\lambda}} = 2$. By Proposition 1.6, $|\cdot|_\lambda$ is a valuation satisfying the triangle inequality.

Proposition 1.11. *For any* x, y *in* $\mathbb{K}$,

$$\left|\left|x\right| - \left|y\right|\right|_\infty \leq \left|x-y\right|,$$

where $|\cdot|_\infty$ *is the absolute value on* $\mathbb{R}$

Proof. $|x| = |(x-y)+y| \leq |x-y| + |y|$ which implies that $|x| - |y| \leq |x-y|$. Similarly $|y| = |(x-y)-x| \leq |x-y| + |x|$ which implies that $|y| - |x| \leq |x-y|$ and therefore,

$$\left|\left|x\right| - \left|y\right|\right|_\infty \leq \left|x-y\right|.$$

Definition 1.12. A valuation $|\cdot|$ on $\mathbb{K}$ satisfies the *ultrametric inequality* if for any x, y in $\mathbb{K}$

$$\left|x+y\right| \leq \max\left\{\left|x\right|, \left|y\right|\right\}.$$

Proposition 1.13. *A valuation* $|\cdot|$ *on* $\mathbb{K}$ *satisfies the ultrametric inequality if and only if one can take* $C = 1$ *in (3) of Definition 1.1*

Proof. Suppose that one can take $C = 1$ in (3) of Definition 1.1. Let x, y, which we may assume to be non-zero, be in $\mathbb{K}$ and suppose that $|x| \leq |y|$ then

$$\left|\frac{x}{y}\right| \leq 1 \text{ and therefore } \left|\frac{x}{y} + 1\right| \leq 1, \text{ hence } \left|x + y\right| \leq \left|y\right| = \max\left\{\left|x\right|, \left|y\right|\right\}.$$

The case $|y| \leq |x|$ is handled similarly. We may conclude that the valuation satisfies the ultrametric inequality.

Next suppose that the valuation satisfies the ultrametric inequality and let x be in $\mathbb{K}$ such that $|x| \leq 1$, then

$$\left|x + 1\right| \leq \max\left\{\left|x\right|, \left|1\right|\right\} = 1.$$

Therefore, one can take $C = 1$ in (3) of Definition 1.1.

Definition 1.14. A valuation on $\mathbb{K}$ is called *non-archimedean* if it satisfies the ultrametric inequality.

From Definition 1.14 it follows that a valuation is *archimedean* if it is not *non-archimedean*.

Proposition 1.15. *Every valuation on $\mathbb{K}$ that is equivalent to a non-archimedean valuation is itself non-archimedean.*

Proof. Suppose $|\cdot|_2 = |\cdot|_1^\lambda$ for some positive real number λ and suppose that $|\cdot|_1$ is non-archimedean. Suppose $|x|_2 \leq 1$, then $|x|_1^\lambda \leq 1$, $|x|_1 \leq 1$, $|x + 1|_1 \leq 1$, $|x + 1|_1^\lambda \leq 1$, $|x + 1|_2 \leq 1$. Therefore $|\cdot|_2$ is non-archimedean.

The following result is very useful:

Proposition 1.16. *Let $|\cdot|$ be a non-archimedean valuation on $\mathbb{K}$. Let x, y be in $\mathbb{K}$ such that $|x| < |y|$, then*

$$\left|x + y\right| = \left|y\right|.$$

Proof. First $|x+y| \leq |y|$, next, $|y| = |(x+y)-x| \leq \max(|x+y|, |x|)$. If $|x+y| < |x|$ then we would have $|y| \leq |x|$ which is against our assumption, therefore $|x+y| \geq |x|$ and hence $|y| \leq |x + y|$. We can conclude that $|x + y| = |y|$.

1.1.2 The Topology Induced by a Valuation on $\mathbb{K}$

Let $|\cdot|$ be a valuation on $\mathbb{K}$, then by Proposition 1.10, we may, and will, assume that it satisfies the triangle inequality. It induces a natural distance function d on $\mathbb{K}$ giving it a structure of a metric space $(\mathbb{K}, d)$.

Proposition 1.17. *Let* $d : \mathbb{K} \times \mathbb{K} \to \mathbb{R}_+$ *be defined by*

$$d(x, y) = \left| x - y \right|$$

then, d is a distance function on $\mathbb{K}$ and $(\mathbb{K}, d)$ is a metric space.

Proof. Suppose $d(x, y) = 0$, then $|x - y| = 0$ and by 1.1 (a), $x = y$. From (4) of Proposition 1.2, $d(x, y) = d(y, x)$ for all x, y in $\mathbb{K}$. For any x, y, z in $\mathbb{K}$,

$$\begin{aligned}
d(x, z) &= \left| x - z \right| \\
&= \left| (x - y) + (y - z) \right| \\
&\leq \left| x - y \right| + \left| y - z \right| \\
&= d(x, y) + d(y, z)
\end{aligned}$$

and hence $(\mathbb{K}, d)$ is a metric space.

Corollary 1.18. *For any x, y, z in $\mathbb{K}$,*

$$\left| d(x, z) - d(y, z) \right|_\infty \leq d(x, y).$$

Since $\mathbb{K}$ is a metric space, the fundamental system of neighborhoods of every element a in $\mathbb{K}$ consists of the open balls of the form:

$$B(a, r) = \left\{ x \in \mathbb{K} : \left| x - a \right| < r \right\},$$

where r is a positive real number.

It is remarkable that any open ball $B(a, R)$ is such that any element in it is its center, in other words, for any $b \in B(a, r)$, $B(b, r) = B(a, r)$.

Proposition 1.19. *Equivalent valuations induce the same topology on $\mathbb{K}$.*

Proof. Suppose $| \cdot |$ and $| \cdot |_\lambda = | \cdot |^\lambda$, with λ a positive real number, are equivalent valuations on $\mathbb{K}$. Let a be in $\mathbb{K}$ and for any $\epsilon > 0$, let $B(a, \epsilon)$ be the open ball associated with $| \cdot |$ and $B_\lambda(a, \epsilon)$ the open ball associated with $| \cdot |_\lambda$. Then for any positive real number r

$$B(a, r) \subset B_\lambda(a, r^\lambda) \quad B_\lambda(a, r) \subset B(a, r^{\frac{1}{\lambda}}).$$

Therefore the two valuations induce the same topology on $\mathbb{K}$.

Actually, the converse is also true. More precisely, we have:

Proposition 1.20. *Let $|\cdot|_1$ and $\cdot|_2$ be two non-trivial valuations which induce the same topology on $\mathbb{K}$, then they are equivalent.*

We begin with a lemma.

Lemma 1.21. *For any x in $\mathbb{K}$, $|x|_1 < 1$ implies $|x|_2 < 1$.*

Proof. Suppose $|x|_1 < 1$ then $|x|_1^n \to 0$ as $n \to \infty$. Since the two valuations induce the same topology, $|x|_2^n \to 0$ as $n \to \infty$, and this implies that $|x|_2 < 1$.

Lemma 1.22. *Suppose that for any x in $\mathbb{K}$, $|x|_1 < 1$ implies $|x|_2 < 1$, then $|\cdot|_1$ and $|\cdot|_2$ are equivalent.*

Proof. Using x^{-1} it is clear that $|x|_1 > 1$ implies $|x|_2 > 1$. Suppose now that there exists x such that $|x|_1 = 1$ but $|x|_2 \neq 1$. Replacing x by x^{-1} if necessary, we may assume that $|x|_2 > 1$. Since $|\cdot|_1$ is non-trivial, there exists y such that $|y|_1 < 1$, which implies that $|y|_2 < 1$. For any positive integer n, consider $z = yx^n$, then, $|z|_1 = |y|_1|x|_1^n = |y|_1 < 1$ which, by hypothesis, implies that $|z|_2 < 1$. However, $|z|_2 = |y|_2|x|_2^n > 1$ for n sufficiently large. This contradiction shows that $|x|_1 = 1$ implies that $|x|_2 = 1$.

The situation is then as follows:

$$|x|_1 < 1 \ \text{ implies } \ |x|_2 < 1,$$
$$|x|_1 > 1 \ \text{ implies } \ |x|_2 > 1,$$
$$|x|_1 = 1 \ \text{ implies } \ |x|_2 = 1.$$

Let u, v be in $\mathbb{K}$ with $|v|_1 \neq 1$, which implies that $|v|_2 \neq 1$. For any integers m, n put $x = u^m v^n$, then after taking logarithms, we obtain

$$m \log |u|_1 + n \log |v|_1 < 0 \ \text{ implies } \ m \log |u|_2 + n \log |v|_2 < 0,$$
$$m \log |u|_1 + n \log |v|_1 > 0 \ \text{ implies } \ m \log |u|_2 + n \log |v|_2 > 0,$$
$$m \log |u|_1 + n \log |v|_1 = 0 \ \text{ implies } \ m \log |u|_2 + n \log |v|_2 = 0.$$

It now follows that

$$\log |u|_1 = \frac{\log |v|_1}{\log |v|_2} \log |u|_2 .$$

Put $\lambda = \frac{\log |v|_1}{\log |v|_2}$, then $\lambda > 0$ and

$$|u|_1 = |u|_2^{\lambda} .$$

Since this is true for any u in $\mathbb{K}$, it follows that $|\cdot|_1$ and $|\cdot|_2$ are equivalent.

The topology on $\mathbb{K}$ turns it into a topological field, in other words, the field operations on $\mathbb{K}$ are continuous. Now that a metric space structure exists on $\mathbb{K}$, one can construct its completion using Cauchy sequences.

Definition 1.23. Let $|\cdot|$ be a valuation on $\mathbb{K}$. A *completion* of $\mathbb{K}$ is a field $\mathbb{F}$ containing $\mathbb{K}$ together with a valuation $||\cdot||$ on it, such that:

(a) $\mathbb{F}$ is a complete metric space with respect to the distance induced by $||\cdot||$;
(b) the valuation $||\cdot||$ extends $|\cdot|$, meaning that for any x in $\mathbb{K}$, $||x|| = |x|$; and
(c) $\mathbb{F}$ is the closure of $\mathbb{K}$ with respect to the topology induced by $||\cdot||$.

The following theorem holds and we refer to [10] for its proof.

Theorem 1.24. *Let $\mathbb{K}$ be a field with a valuation $|\cdot|$. A completion exists and any two completions are canonically isomorphic.*

In the later parts of the book we will exclusively use a field with a valuation with respect to which it is complete.

1.1.3 Non-Archimedean Valuations

All valuations under consideration here will be non-trivial.

Definition 1.25. Let $\mathbb{K}$ be a field with a non-archimedean valuation $|\cdot|$, then

$$A = \left\{ x \in \mathbb{K} : \left| x \right| \leq 1 \right\}$$

is called the *valuation ring* (or the *ring of integers*) of $\mathbb{K}$.

The following proposition is easy to prove.

Proposition 1.26. *The following hold:*

(a) A is a local ring;
(b) $U = \{x \in A : |x| = 1\}$ is the group of units in A; and
(c) $M = \{x \in A : |x| < 1\}$ is the unique maximal ideal of A.

Definition 1.27. The *value group* of $\mathbb{K}$ is the image of $\mathbb{K}^*$ under the valuation map $|\cdot|$. It is denoted $|\mathbb{K}^*|$.

The value group $|\mathbb{K}^*|$ is a multiplicative group of positive real numbers, hence it is either:

(a) everywhere dense, or
(b) infinite cyclic.

Definition 1.28. In the case where the value group is infinite cyclic, the valuation is called a *discrete valuation* and in the case where the value group is everywhere dense, the valuation is called a *dense valuation*.

Proposition 1.29. *The valuation* $|\cdot|$ *is a discrete valuation on* $\mathbb{K}$, *if and only if* M *is a principal ideal.*

Suppose the valuation is discrete, hence the value group $|\mathbb{K}^*|$ is infinite cyclic. Let ξ be in $\mathbb{K}^*$ such that $|\xi|$ generates $|\mathbb{K}^*|$. Clearly $|\xi| \neq 1$. Every element x in $\mathbb{K}^*$ is of the form $x = u.\xi^n$ for some unit u and some integer n. Let π be either ξ or ξ^{-1} but so that $|\pi| < 1$, therefore $\pi \in M$. Now it is clear that π generates the maximal ideal M. Conversely suppose M is a principal ideal generated by π.

Lemma 1.30. *If* $x \in \mathbb{K}$, *then, there exists an integer* n *in* $\mathbb{Z}$ *and a unit* u *in* U *such that* $x = u.\pi^n$.

Proof. If x is in U, then x is a unit and $x = x.\pi^0$. Next suppose x is in M so that $x = \pi.y_1$ with $y_1 \in A$. If $y_1 \in U$, then we are done, if not $y_1 \in M$ and $y_1 = \pi.y_2$ with $y_2 \in A$ and $x = \pi^2.y_2$, and so on, there exists a sequence $\{y_k\}$ in A such that $x = \pi^k.y_k$ for all k. We claim that the sequence stops. Indeed if it does not, then there exists a strictly increasing sequence of positive real numbers $\{|y_k|\}$ which is also bounded since $|y_k| \leq 1$ for all k. The sequence $\{|y_k|\}$ therefore converges. However, since $y_k = \pi^{-k}.x$, we see that $\{|y_k|\}$ diverges. This contradiction shows that the sequence stops, and there exists n such that $x = \pi^n.y_n$ with $y_n \in U$.

Finally, suppose x is in $\mathbb{K} - A$ then $|x| > 1$ and hence $x^{-1} \in M$ which implies that $x = \pi^n.u$ for some negative integer n and some unit u.

It is now clear that the value group $|\mathbb{K}^*|$ is a cyclic subgroup of $\mathbb{R}_+^*$ generated by $|\pi|$, and hence, the valuation is discrete.

There is another criterion for a non-archimedean valuation to be discrete, with a more topological flavor.

Proposition 1.31. *The non-archimedean valuation* $|\cdot|$ *on* $\mathbb{K}$ *is discrete if and only if* 1 *is an isolated point in* $|\mathbb{K}^*|$.

Here 1 is an isolated point in $|\mathbb{K}^*|$ means that there exists an open interval centered at 1 such that the intersection of this open interval with $|\mathbb{K}^*|$ is reduced to $\{1\}$. In other words there exists $\delta > 0$ such that

$$\left(1 - \delta, 1 + \delta\right) \cap \left|\mathbb{K}^*\right| = \left\{1\right\}.$$

Suppose the valuation is discrete. Let c be a generator for $|\mathbb{K}^*|$. We can choose $c > 1$ and $|\mathbb{K}^*| = \{c^n : n \in \mathbb{Z}\}$. Let δ be so chosen that

$$0 < \delta < \min(1, c - 1).$$

Then if,

$$1 - \delta < c^n < 1 + \delta,$$

then

$$\log_c(1 - \delta) < n < \log_c(1 + \delta).$$

By the choice of δ, the following inequalities hold

$$\log_c(1 - \delta) < 0 < \log_c(1 + \delta) < 1$$

this forces $n = 0$ and hence 1 is an isolated point in $|\mathbb{K}^*|$.

Suppose now that 1 is an isolated point in $|\mathbb{K}^*|$. This means that there exists $\delta > 0$, which we may assume to be less than 1, such that

$$x \in \mathbb{K}^* \text{ and } 1 - \delta < |x| < 1 + \delta \text{ imply } |x| = 1.$$

Lemma 1.32. *For any $x \in \mathbb{K}^*$, $|x|$ is an isolated point in $|\mathbb{K}^*|$.*

Proof. Let $x \in \mathbb{K}^*$ and suppose $y \in \mathbb{K}$ with

$$\left| x \right| - \delta \left| x \right| < \left| y \right| < \left| x \right| + \delta \left| x \right|$$

then, multiplying by $|x|^{-1}$,

$$1 - \delta < \left| \frac{y}{x} \right| < 1 + \delta$$

which implies

$$\left| \frac{y}{x} \right| = 1, \quad \text{and} \quad |y| = |x|.$$

Therefore

$$\left| \mathbb{K}^* \right| \cap \left(\left| x \right| - \delta \left| x \right|, \left| x \right| + \delta \left| x \right| \right) = \left\{ \left| x \right| \right\}$$

and $|x|$ is an isolated point.

Lemma 1.33. *Let $S = |\mathbb{K}^*| \cap (0, 1)$ and $s = \sup S$, then $s < 1$ and $s \in S$.*

Proof. Since S is bounded, $s = \sup S$ exists and $s \leq 1$. We must have $s < 1$. Otherwise, $s = 1$, but this would contradict the facts that $s = \sup S$, $1 \notin S$ and 1 is an isolated point. Therefore $s < 1$.

Suppose there is a strictly increasing sequence $\{|x_n|\}_n$ in S, with $x_n \in \mathbb{K}^*$, such that $\lim_{n \to \infty} |x_n| = s$, then there exists N such that for all $n \geq N$,

$$\left| x_n \right| \in \left(\frac{s}{1 + \delta}, s \right].$$

On the one hand, the inequality

$$\frac{s}{1+\delta} < \left|x_n\right| \le s$$

implies

$$s < \left|x_n\right| + \delta\left|x_n\right|$$

and on the other hand, the inequalities

$$\left|x_n\right| \le s < \frac{s}{1-\delta}$$

imply

$$\left|x_n\right| - \delta\left|x_n\right| < s.$$

Therefore, as $1 - \delta > 0$, for all $n \ge N$ we have

$$\left|x_N\right| - \delta\left|x_N\right| < \left|x_n\right| - \delta\left|x_n\right| < s < \left|x_N\right| + \delta\left|x_N\right| < \left|x_n\right| + \delta'\left|x_n\right|.$$

These inequalities, together with

$$\left|x_N\right| < \left|x_n\right| \le s, \quad \left|x_n\right| \notin \left(\left|x_N\right| - \delta\left|x_N\right|, \left|x_N\right| + \delta\left|x_N\right|\right) \text{ (proof of Lemma 1.33)}$$

yield a contradiction to the assumption that there exists a strictly increasing sequence in S converging to s. Therefore, any sequence converging to s is stationary. This implies that $s \in S$, and $s = |\pi|$ for some $\pi \in \mathbb{K}^*$.

To finish the proof of Proposition 1.31, we claim that

$$\forall x \in \mathbb{K}^*, \text{ there exists } n \in \mathbb{Z} \text{ such that } \left|x\right| = s^n.$$

Clearly, it is enough to prove this for $x \in \mathbb{K}^*$ such that $|x| \in S$.

Let $|x| \in S$, then $|x| \le s$. If $|x| = s$, then, we are done. If not, $|x| < s$ and hence $\frac{|x|}{s} < 1$. This combined with $s = \pi \in \mathbb{K}^*$, implies that $\frac{|x|}{s}$ actually is in S, we have

$$\frac{|x|}{s} \le s \text{ hence } \left|x\right| \le s^2.$$

If $|x| = s^2$, then, we are done. If not, $|x| < s^2$ and hence $\frac{|x|}{s^2} < 1$. Then, as before, we have

$$\frac{|x|}{s^2} \leq s \text{ hence } |x| \leq s^3.$$

This process stops and eventually we find N such that $|x| = s^N$. Otherwise, if the process continues, then we would have a sequence $\{s^n\}_n$ such that $|x| < s^n$. As $s < 1$, the sequence $\{s^n\}_n$ converges to 0, however $|x| > 0$ and we obtain a contradiction. This concludes the proof of Proposition 1.31.

Definition 1.34. A generator for M is called a *prime* element or a *uniformizer* for the valuation.

We now consider extensions of non-archimedean valued fields.

Proposition 1.35. *Let $\mathbb{F}$ be a field extension of $\mathbb{K}$ and $|\cdot|$ a valuation on $\mathbb{F}$, then, $|\cdot|$ is non-archimedean on $\mathbb{F}$ if and only if it is non-archimedean on $\mathbb{K}$.*

Proof. Without loss of generality, we may assume that the valuation satisfies the triangle inequality. If the valuation is non-archimedean on $\mathbb{F}$, then it is so on $\mathbb{K}$. Now suppose it is non-archimedean on $\mathbb{K}$. This implies that for any positive integer N, $|N| \leq 1$. Next let x be in $\mathbb{F}$ such that $|x| \leq 1$, then, for any positive integer n,

$$
\left| x + 1 \right|^n = \left| (x+1)^n \right|
$$
$$
= \left| \sum_{j=0}^{n} \binom{n}{j} x^j \right|
$$
$$
\leq \sum_{j=0}^{n} \left| \binom{n}{j} x^j \right|
$$
$$
= \sum_{j=0}^{n} \left| \binom{n}{j} \right| |x|^j
$$
$$
\leq \sum_{j=0}^{n} |x|^j
$$
$$
\leq \left(n + 1 \right).
$$

Taking n-th root and letting $n \to \infty$ yields $|x+1| \leq 1$ and $|\cdot|$ is non-archimedean on $\mathbb{F}$.

Proposition 1.36. *The only valuation on a finite field is the trivial valuation.*

Proof. This is so because if q is the order of the finite field, then for any non-zero x, $x^{q-1} = 1$, hence $|x^{q-1}| = 1$ and since $|x| \in \mathbb{R}_+^*$, it follows that $|x| = 1$.

Proposition 1.37. *If* char$(\mathbb{K})$ *is finite, then every valuation on* $\mathbb{K}$ *is non-archimedean.*

Proof. This is so because $\mathbb{K}$ contains a finite field.

Let $\mathbb{K}$ be endowed with a non-archimedean valuation $|\cdot|$ and $\mathbb{F}$ its completion. We also denote by $|\cdot|$ the valuation on $\mathbb{F}$ extending that of $\mathbb{K}$. Let $A_{\mathbb{K}}$ and $M_{\mathbb{K}}$ (resp. $A_{\mathbb{F}}$ and $M_{\mathbb{F}}$) denote the ring of integers and the maximal ideal of $\mathbb{K}$ (resp. $\mathbb{F}$), then we have the following proposition,

Proposition 1.38. *With the notation above*

(a) $\mathbb{K} \cap A_{\mathbb{F}} = A_{\mathbb{K}}$;
(b) $\mathbb{K} \cap M_{\mathbb{F}} = M_{\mathbb{K}}$;
(c) $A_{\mathbb{K}}$ *is dense in* $A_{\mathbb{F}}$; *and*
(d) *The residue class fields* $A_{\mathbb{K}}/M_{\mathbb{K}}$ *and* $A_{\mathbb{F}}/M_{\mathbb{F}}$ *are isomorphic.*

Proof. Statements (a) and (b) are clear. (c) Let x be in $\mathbb{K} \cap A_{\mathbb{F}}$, then, since $\mathbb{K}$ is dense in $\mathbb{F}$, there exists a sequence $\{x_n\}_n$ in $\mathbb{K}$ that converges to x. Hence for n sufficiently large, $|x_n - x| < 1$ but then, for n sufficiently large, $|x_n| \leq 1$, hence all but a finite number of the terms of the sequence is in $A_{\mathbb{K}}$, therefore $A_{\mathbb{K}}$ is dense in $A_{\mathbb{F}}$.

(d) Let $\phi : A_{\mathbb{K}} \to A_{\mathbb{F}}/M_{\mathbb{F}}$ be defined by $\phi(a) = a \mod M_{\mathbb{F}}$. This is well-defined by (a) and it is a ring homomorphism. Now let x be in $A_{\mathbb{F}}$, then, by (c), there exists a in $A_{\mathbb{K}}$ such that $|x - a| < 1$, therefore $\phi(a) = x \mod M_{\mathbb{F}}$ and ϕ is surjective. By (b), ker $\phi = M_{\mathbb{K}}$, hence by the first isomorphism theorem, (d) follows.

1.1.4 Some Analysis on a Complete Non-Archimedean Valued Field

Now we assume that $\mathbb{K}$ is endowed with a non-archimedean valuation $|\cdot|$, under which it is a complete metric space.

Definition 1.39. A series $\sum_{j=0}^{\infty} x_j$, $x_j \in \mathbb{K}$ for all n, *converges* if the sequence of partial sums $\{s_n\} = \left\{\sum_{j=0}^{n} x_j\right\}_n$ converges in $\mathbb{K}$. If $s = \lim s_n$, then we say that $\sum_{j=0}^{\infty} x_j = s$.

Proposition 1.40. *The series* $\sum_{j=0}^{\infty} x_j$ *converges if and only if the sequence* $\{x_j\}_j$ *converges to* 0.

Proof. It is standard that if the series converges then the general term converges to 0. Now suppose $\{x_j\}_j$ converges to 0. For every n let $s_n = \sum_{j=0}^{n} x_j$ be the n-th partial sum. Since $\{x_j\}_j$ converges to 0,

$$\forall \epsilon > 0 \ \exists N \text{ such that for } j \geq N \ \left| x_j \right| < \epsilon.$$

Now let $m > n > N$, then

$$\left| s_m - s_n \right| = \left| \sum_{j=n+1}^{m} x_j \right| \leq \max \left\{ \left| x_j \right| : n + 1 \leq j \leq m \right\} < \epsilon.$$

Therefore the sequence of partial sums $\{s_n\}_n$ is a Cauchy sequence in $\mathbb{K}$, and since $\mathbb{K}$ is complete, the sequence converges, therefore the series converges.

We now assume that $\mathbb{K}$ is complete with respect to a discrete valuation $|\cdot|$. We denote by π be a uniformizer and by S a complete set of representatives for the residue class field A/M.

Proposition 1.41. *The series* $\sum_{j=0}^{\infty} a_j \pi^j$, *where* $a_j \in S$, *converges.*

Proof. $|a_j \pi^j| = |a_j||\pi^j| \leq |\pi^j|$, but $|\pi| < 1$ therefore $|\pi^j| \to 0$ as $j \to \infty$, hence by Proposition 1.37 the series converges.

We have the following important theorem for a field which is complete with respect to a discrete valuation.

Theorem 1.42. *Let* $\mathbb{K}$ *be complete with respect to a discrete valuation* $|\cdot|$. *Let* π *be a uniformizer and* S *a complete set of representatives for the residue class field* A/M. *Then, every* x *in* A *can be written uniquely as an infinite series*

$$x = \sum_{j=0}^{\infty} a_j \pi^j,$$

where $a_j \in S$ *for all* j.

Proof. Let x be in A, then, there exists a_0 in S such that

$$a - a_0 = \pi b_1, \quad b_1 \in A.$$

Since b_1 is in A, there exists a_1 in S such that

$$b_1 - a_1 = \pi b_2, \quad b_2 \in A.$$

Therefore $a = a_0 + \pi a_1 + \pi^2 b_2$. Continuing in this fashion, we obtain for any positive integer n,

$$a = \sum_{j=0}^{n-1} a_j \pi^j + b_n \pi^n$$

with b_n in A. Note that $|b_n \pi^n| \leq |\pi^n|$, $|\pi| < 1$ and therefore, $|b_n \pi^n| \to 0$ as $n \to \infty$. We conclude that

$$a = \sum_{j=0}^{\infty} a_j \pi^j, \quad a_j \in S.$$

Corollary 1.43. *Let $\mathbb{K}$ be complete with respect to a discrete valuation $|\cdot|$. Let π be a uniformizer and S a complete set of representatives for the residue class field A/M. Assume $0 \in S$. Then, every x in $\mathbb{K}^*$ can be written uniquely as an infinite series*

$$x = \sum_{j=-N}^{\infty} a_j \pi^j,$$

where $a_j \in S$ for all j, N is a non-negative integer and $a_{-N} \neq 0$.

Proof. The case x in A is Theorem 1.42, and for that case $N = 0$. Now for x in $\mathbb{K} - A$, $|x| > 1$. Since $|\pi^{-1}| > 1$, there exists a positive integer M such that

$$\left|\pi^{-1}\right|^{M-1} \leq |x| < \left|\pi^{-1}\right|^{M}.$$

Multiplying through by $|\pi|^M$ yields

$$|\pi| \leq \left|\pi^M x\right| < 1.$$

Therefore, by Theorem 1.39, and with a suitable power of π and an adjustment of the indices,

$$x = \sum_{j=-N}^{\infty} a_j \pi^j$$

for some positive integer N, $a_j \in S$ and $a_{-N} \neq 0$.

1.1.5 The Order Function for a Discrete Valuation

Our approach to valuation theory is multiplicative, in analogy with the usual absolute value on $\mathbb{C}$ or $\mathbb{R}$. There is an additive approach which we shall describe briefly in this subsection. We shall define the *additive valuation*, also called the *order function* for a discrete valuation. We shall use this approach only occasionally in later parts of the book and focus, instead, on the multiplicative approach which is more convenient for use in non-archimedean Banach spaces and Operator Theory. In this subsection we assume that $|\cdot|$ is a non-archimedean discrete valuation on $\mathbb{K}$. In this situation the value group $|\mathbb{K}^*|$ is infinite cyclic.

Definition 1.44. An *additive valuation* or an *order function*, denoted ord is a function ord : $\mathbb{K} \to \mathbb{Z} \cup \{\infty\}$ satisfying

(a) $\mathrm{ord}(0) = \infty$ by convention;
(b) $\mathrm{ord}(xy) = \mathrm{ord}(x) + \mathrm{ord}(y)$ for any non-zero x, y in $\mathbb{K}$; and
(c) $\mathrm{ord}(x + y) \geq \min(\mathrm{ord}(x), \mathrm{ord}(y))$ for any non-zero x, y in $\mathbb{K}$.

The relationship between the two approaches is now described. Let π be a uniformizer for the valuation $|\cdot|$, then:

(a) For any non-zero x in $\mathbb{K}$, $x = \pi^{\mathrm{ord}(x)}$;
(b) The ring of integers $A = \{x \in \mathbb{K} : |x| \leq 1\} = \{x \in \mathbb{K} : \mathrm{ord}(x) \geq 0\}$;
(c) The maximal ideal $M = \{x \in A : |x| < 1\} = \{x \in A : \mathrm{ord}(x) > 0\}$.

Proposition 1.45. *For any non-zero x, y in $\mathbb{K}$, if $\mathrm{ord}(x) \neq \mathrm{ord}(y)$, then*

$$\mathrm{ord}\left(x + y\right) = \min\left(\mathrm{ord}(x), \mathrm{ord}(y)\right).$$

Proposition 1.46. *Let $\mathbb{K}$ be complete with respect to a discrete valuation $|\cdot|$. Let S be a complete set of representatives for the residue class field and π a uniformizer.*

If x is in $\mathbb{K}$ and is written as $\displaystyle\sum_{j=N}^{\infty} a_j \pi^j$ where $N \in \mathbb{Z}$ and $a_j \in S$ for all j,

then $\mathrm{ord}(x) = N$.

1.2 Examples

In this section we illustrate the theory developed in Sect. 1.1 with many examples.

1.2.1 Examples of Archimedean Valuation

The ordinary absolute value on $\mathbb{C}$, on $\mathbb{R}$ and on any subfield, is the typical example of archimedean valuations. In fact one can prove the following Theorem (see Cassels [10]).

Theorem 1.47. *Let $\mathbb{K}$ be complete with respect to an archimedean valuation $|\cdot|$, then $\mathbb{K}$ is isomorphic to either $\mathbb{R}$ or $\mathbb{C}$, and $|\cdot|$ is equivalent to the ordinary absolute value.*

1.2.2 Examples of Non-Archimedean Valued Fields

In this subsection and for the rest of this chapter, we focus on non-archimedean valuations.

Example 1.48 (The field $\mathbb{Q}$ of rational numbers). This is a classic example and we will work out the details. Let p be a prime number, then, because of the unique factorization in $\mathbb{Z}$, every non-zero rational number x can be written as

$$x = \frac{a}{b} p^n$$

where n, a, b are integers, and $\gcd(p, ab) = 1$.

Put

$$\left| x \right|_p = p^{-n} \; \text{ if } \; x \neq 0 \; \text{ and } \; \left| 0 \right|_p = 0.$$

Then we have the following

Proposition 1.49. $|\cdot|_p$ *is a valuation on $\mathbb{Q}$, called the p-adic valuation.*

Proof. From the definition $|x|_p = 0$ if and only if $x = 0$. If $x = p^n \frac{a}{b}$ and $y = p^m \frac{c}{d}$ then $xy = p^{n+m} \frac{ac}{bd}$, $\gcd(p, abcd) = 1$, therefore $|xy|_p = p^{-(n+m)} = |x|_p |y|_p$. If $n \leq m$ then $x + y = p^n (\frac{a + p^{m-n} c}{bd})$ and hence $|x + y| \leq p^{-n} = \max\{|x|_p, |y|_p\}$. The case $m \leq n$ is handled similarly.

It is useful to also use the additive valuation, or order function in this case. The order function is denoted ord_p. The relationship between the two approaches is: for all $x \in \mathbb{Q}$,

$$\left| x \right|_p = p^{-\mathrm{ord}_p(x)}.$$

Proposition 1.50. *The following hold:*

(a) $\mathrm{ord}_p(x) = \infty$ *if and only if $x = 0$;*
(b) $\mathrm{ord}_p(xy) = \mathrm{ord}_p(x) + \mathrm{ord}_p(y)$;
(c) $\mathrm{ord}_p(x + y) \geq \min\{\mathrm{ord}_p(x), \mathrm{ord}_p(y)\}$.

Proposition 1.51. *The p-adic valuation is a discrete valuation.*

Proof. $|\mathbb{Q}^*| = \{p^n : n \in \mathbb{Z}\} = p^{\mathbb{Z}}$, hence it is an infinite cyclic multiplicative subgroup of the group $\mathbb{R}_+^*$.

(a) The ring valuation ring is $\mathbb{Z}_{(p)} = \{x \in \mathbb{Q} : \mathrm{ord}_p(x) \geq 0\}$.
(b) The unique maximal ideal is $M_{(p)} = \{x \in \mathbb{Q} : \mathrm{ord}_p(x) > 0\} = p\mathbb{Z}_{(p)}$. p is a uniformizer.
(c) The group of units is $U_{(p)} = \{x \in \mathbb{Q} : \mathrm{ord}_p(x) = 0\}$.

Proposition 1.52. *The residue class field $\mathbb{Z}_{(p)}/M_{(p)}$ is isomorphic to $\mathbb{Z}/p\mathbb{Z}$.*

Proof. Let $\phi : \mathbb{Z} \to \mathbb{Z}_{(p)}/M_{(p)}$ be defined by

$$\phi(n) = n \mod M_{(p)}.$$

This map is well-defined since $\mathbb{Z} \subset \mathbb{Z}_{(p)}$ and it is a homomorphism of rings. Moreover, it is surjective, because $\phi(0) = 0 \mod M_{(p)}$ and if $a \mod M_{(p)} \neq 0$ then $\text{ord}_p(a) = 0$ hence there exists $n \in \mathbb{Z}$ such that $n - a \in M_{(p)}$ and $\phi(n) = a \mod M_{(p)}$. The kernel of ϕ is $\mathbb{Z} \cap M_{(p)} = M_p = p\mathbb{Z}$. Now the first isomorphism theorem yields the result.

(a) The completion of $\mathbb{Q}$ with respect to $|\cdot|_p$ is called the field of p-adic numbers and is denoted $\mathbb{Q}_p$. Without loss of generality, the valuation on the completion, $\mathbb{Q}_p$, which extends the p-adic valuation on $\mathbb{Q}$ is still denoted $|\cdot|_p$. (b) The ring of integers, which is $\{x \in \mathbb{Q}_p : |x|_p \leq 1\}$ is denoted $\mathbb{Z}_p$. (c) The unique maximal ideal, which is $\{x \in \mathbb{Z}_p : |x|_p < 1\} = p\mathbb{Z}_p$. (d) The prime p is a uniformizer. (e) The residue class field is $\mathbb{Z}_p/p\mathbb{Z}_p \cong \mathbb{Z}/p\mathbb{Z}$. (f) Every p-adic number x can be expressed as an infinite series $x = \sum_{j=N}^{\infty} a_j p^j$ with $a_j \in \{0, 1, 2, \ldots, p-1\}$ and $N \in \mathbb{Z}$.

Example 1.53 (The field $F(T)$ of rational functions with coefficients in a field F (Part 1)). Let F be a field and $F[T]$ the ring of polynomials in the variable T with coefficients in F. The quotient field of $F[T]$ is the field $F(T)$ of rational functions. The elements of $F(T)$ are quotients of polynomials. Let $P(T)$ be a monic irreducible polynomial in $F[T]$. Since $F[T]$ is a PID, every polynomial can be (essentially) uniquely written as a product of irreducible polynomials. As a consequence of this fact, every element $f(T)$ of $F(T)$ can be written as

$$f(T) = (P(T))^n \frac{Q_1(T)}{Q_2(T)},$$

where $Q_1(T)$ and $Q_2(T)$ are polynomials in $F[T]$, such that $P(T)$ does not divide the product $Q_1(T)Q_2(T)$ and $n \in \mathbb{Z}$.

Proposition 1.54. *In the notation above, let*

$$\text{ord}_{P(T)} : F(T) \to \mathbb{Z} \cup \{\infty\}$$

be defined by

$$\text{ord}_{P(T)}(f(T)) = n, \quad \text{ord}_{P(T)}(0) = \infty.$$

Then $\text{ord}_{P(T)}$ is an order function on $F(T)$.

Proposition 1.55. *Let c be a real number satisfying $0 < c < 1$. In the above notation, let $|\cdot|_{P(T)} : F(T) \to \mathbb{R}_+^*$ be defined by,*

$$\left| f(T) \right|_{P(T)} = c^n, \quad \left| 0 \right| = 0$$

then $| \cdot |_{P(T)}$ is a discrete valuation.

This is the multiplicative version of Proposition 1.54.

Let us specialize furthermore by considering the classical case of $P(T) = T$ to illustrate the theory.

Proposition 1.56. *The following hold:*

(a) *The ring of integers, denoted A_T, is*

$$\left\{ \frac{P(T)}{Q(T)} : P(T), Q(T) \in F[T], T \text{ does not divide } Q(T) \right\} ;$$

(b) *The maximal ideal, denoted M_T, is*

$$\left\{ \frac{P(T)}{Q(T)} : P(T), Q(T) \in F[T], T \text{ divides } P(T), T \text{ does not divide } Q(T) \right\} ;$$

(c) *T is a uniformizer;*

(d) *The residue class field A_T/M_T is isomorphic to F;*

(e) *The completion of $F(T)$ is the field $F((T))$ of Laurent series in T. Every element $f(T) \in F((T))$ can be written as*

$$f(T) = \sum_{j=N}^{\infty} a_j T^j,$$

where $N \in \mathbb{Z}$ and $a_j \in F$ for all j.

Proof. Statements (a)–(c) are clear. To prove (d) we consider the map $\phi : A_T \to F$ defined by

$$\phi(f(T)) = f(0).$$

This is well-defined since the denominator of $f(T)$ is prime to T. Moreover, ϕ is a ring homomorphism which is clearly surjective. The kernel of ϕ is the set of $f(T) \in A_T$ whose numerator vanishes at 0, hence T divides the numerator. Therefore, $Ker\phi = M_T$. The first isomorphism theorem yields (d). Statement (e) is also clear.

Remark 1.57. The ring of integers in the completion $F((T))$ is just $F[[T]]$ the ring of formal power series with coefficients in F.

Example 1.58 (The field $F(T)$ of rational functions with coefficients in a field F (Part 2)). Every $f(T) \in F(T)$ can be written as $f(T) = \frac{P(T)}{Q(T)}$ where $P(T)$ and $Q(T)$ are polynomials.

Proposition 1.59. *Let c be a real number with $0 < c < 1$. Define $|\cdot|_\infty : F(T) \to \mathbb{R}_+^*$ by*

$$\left|f(T)\right|_\infty = c^{\deg Q(T) - \deg P(T)}, \quad \left|0\right|_\infty = 0$$

then, $|\cdot|_\infty$ is a discrete valuation on $F(T)$ and the associated order function is

$$\mathrm{ord}_\infty(f(T)) = \deg Q(T) - \deg P(T), \quad \mathrm{ord}_\infty(0) = \infty.$$

Proof. $|f(T)|_\infty = 0$ only if $f(T) = 0$. Let $f_i(T) = \frac{P_i(T)}{Q_i(T)}$ with $i = 1, 2$, then

$$
\begin{aligned}
\left|f_1(T) + f_2(T)\right|_\infty &= \left|\frac{P_1(T)}{Q_1(T)} + \frac{P_2(T)}{Q_2(T)}\right|_\infty \\
&= \left|\frac{P_1(T)Q_2(T) + Q_1(T)P_2(T)}{Q_1(T)Q_2(T)}\right|_\infty.
\end{aligned}
$$

Consider the case $|f_1(T)|_\infty \le |f_2(T)|_\infty$, then,

$$\deg Q_1(T) - \deg P_1(T) \ge \deg Q_2(T) - \deg P_2(T)$$

hence

$$\deg Q_1(T)P_2(T) \ge \deg P_1(T)Q_2(T)$$

and therefore

$$\deg(P_1(T)Q_2(T) + Q_1(T)P_2(T)) \le \deg Q_1(T)P_2(T)$$

which implies

$$\deg Q_1(T)Q_2(T) - \deg(P_1(T)Q_2(T) + Q_1(T)P_2(T)) \ge \deg P_2(T) - \deg P_2(T)$$

and we may conclude that

$$\left|f_1(T) + f_2(T)\right|_\infty \le \left|f_2(T)\right|_\infty.$$

The other case is handled similarly.

Proposition 1.60. *For the valuation $|\cdot|_\infty$, the following hold:*

(a) The ring of integers is

$$A_\infty = \left\{\frac{P(T)}{Q(T)} : P(T), Q(T) \in F[T], \deg P(T) \le \deg Q(T)\right\};$$

(b) The maximal ideal is

$$M_\infty = \left\{ \frac{P(T)}{Q(T)} : P(T), Q(T) \in F[T], \deg P(T) < \deg Q(T) \right\};$$

(c) $\frac{1}{T}$ is a uniformizer;
(d) The residue class field A_∞/M_∞ is isomorphic to F;
(e) The completion of $F(T)$ with respect to $|\cdot|_\infty$ is $F((\frac{1}{T}))$, the Laurent series in $\frac{1}{T}$.

Proof. Statements (a) and (b) are clear in view of the definition of $|\cdot|_\infty$. For (c) let $\frac{P(T)}{Q(T)} \in M_\infty$ then $\deg P(T) < \deg Q(T)$. Say $\deg P(T) = n$ and $\deg Q(T) = m$ with $n < m$. Let

$$P(T) = a_n T^n + \ldots + a_0, \quad Q(T) = b_m T^m + \ldots + b_0, \quad a_i, b_i \in F$$

then,

$$\frac{P(T)}{Q(T)} = \frac{1}{T^{m-n}} \frac{a_n T^m + \ldots + a_0 T^{m-n}}{b_m T^m + \ldots + b_0} \in \frac{1}{T^{m-n}} . A_\infty$$

hence $\frac{1}{T^{m-n}}$ is a uniformizer.

For (d), consider the map $\phi : A_\infty \to F$ defined in the following way: Let $\frac{P(T)}{Q(T)} \in A_\infty$ such that

$$P(T) = a_n T^n + \ldots + a_0, \quad Q(T) = b_m T^m + \ldots + b_0, \quad a_i, b_i \in F$$

then

$$\phi\left(\frac{P(T)}{Q(T)} \right) = \begin{cases} 0 & n < m, \\ \frac{a_n}{b_n} & n = m. \end{cases}$$

The map ϕ is a surjective homomorphism of rings whose kernel is precisely M_∞, and again, we apply the first isomorphism theorem. Statement (e) is also clear from the general theory.

Example 1.61 (Locally compact non-archimedean valued fields (Part 1)). Let L be an algebraic number field. It is a finite extension of $\mathbb{Q}$. An element $z \in L$ is called an integer in L if it is an algebraic integer, in other words, if it is a root of a monic polynomial with coefficient in $\mathbb{Z}$. The set of integers of L is a subring of L and is denoted $\mathcal{O}$, it is a Dedekind domain, meaning, an integral domain in which every non-zero proper ideal is, essentially, a product of prime ideals. The field L is the quotient field of $\mathcal{O}$.

Let $\mathfrak{p}$ be a prime ideal in $\mathcal{O}$, then:

(a) $\mathfrak{p} \cap \mathbb{Z} = p\mathbb{Z}$, p prime in $\mathbb{Z}$;
(b) $\mathcal{O}/\mathfrak{p}$ is a finite field containing $\mathbb{Z}/p\mathbb{Z}$, its order is q a power of p.

Using $\mathfrak{p}$ we are going to construct a discrete valuation on L.

Let

$$A_{\mathfrak{p}} = \left\{ \frac{a}{s} : a \in \mathcal{O}, \ s \notin \mathfrak{p} \right\}.$$

Then, $A_{\mathfrak{p}}$ is a principal ideal domain satisfying the following properties:

(1) $\mathcal{O} \subset A_{\mathfrak{p}} \subset L$;

(2) $A_{\mathfrak{p}}$ is a local ring with maximal ideal $M_{\mathfrak{p}} = \left\{ \frac{a}{s} : a \in \mathfrak{p}, \ s \notin \mathfrak{p} \right\}$ and group of
    units $U_{\mathfrak{p}} = \left\{ \frac{a}{s} : a, s \notin \mathfrak{p} \right\}$;

(3) Every element $x \in A_{\mathfrak{p}}$ can be written, uniquely as $x = \pi^n u, \ n \in \mathbb{N}, u$ a unit in
    $A_{\mathfrak{p}}$ where π is a generator of $M_{\mathfrak{p}}$; and

(4) The quotient field of $A_{\mathfrak{p}}$ is equal to L.

In short $A_{\mathfrak{p}}$ is a discrete valuation ring.

Items (3) and (4) above, imply that every $x \in L$ can be written uniquely as

$$x = \pi^n u$$

where $n \in \mathbb{Z}$, u is a unit in $A_{\mathfrak{p}}$.

In the notations above, define $| \cdot |_{\mathfrak{p}} : L \to \mathbb{R}_+^*$ by

$$|x|_{\mathfrak{p}} = p^{-n}, \ x \neq 0, \ |0|_{\mathfrak{p}} = 0$$

and $\mathrm{ord}_{\mathfrak{p}} : L \to \mathbb{Z} \cup \{\infty\}$ by

$$\mathrm{ord}_{\mathfrak{p}}(x) = n, \ x \neq 0, \ \mathrm{ord}_{\mathfrak{p}}(0) = \infty.$$

The proofs of the next two propositions are straightforward.

Proposition 1.62. $| \cdot |_{\mathfrak{p}} : L \to \mathbb{R}_+^*$ *is a discrete valuation on L and $\mathrm{ord}_{\mathfrak{p}}$ is the associated order function*

Proposition 1.63. *We have the following:*

(a) The valuation ring is

$$A_L = \left\{ \frac{x}{y} : x, y \in \mathcal{O}, \ y \neq 0, \ \mathrm{ord}_{\mathfrak{p}}(x) \geq \mathrm{ord}_{\mathfrak{p}}(y) \right\} = A_{\mathfrak{p}}.$$

(b) The maximal ideal is

$$M_L = \left\{ \frac{x}{y} : x, y \in \mathcal{O}, \ y \neq 0 \ \mathrm{ord}_{\mathfrak{p}}(x) > \mathrm{ord}_{\mathfrak{p}}(y) \right\} = M_{\mathfrak{p}}.$$

(c) A uniformizer is an element $\theta \in L$ such that $\mathrm{ord}_{\mathfrak{p}}(\theta) = 1$.

(d) The group of units

$$U_L = \left\{ \frac{x}{y} : x, y \in \mathcal{O}, \, y \neq 0 \operatorname{ord}_\mathfrak{p}(x) = \operatorname{ord}_\mathfrak{p}(y) \right\} = U_\mathfrak{p}.$$

(e) The residue class field $A_L/M_L = A_\mathfrak{p}/M_\mathfrak{p}$ is isomorphic to $\mathcal{O}/\mathfrak{p}$.

Let $L_\mathfrak{p}$ be the completion of L with respect to $|\cdot|_\mathfrak{p}$, $\mathcal{O}_\mathfrak{p}$ its valuation ring, $\mathfrak{M}_\mathfrak{p}$ its maximal ideal, and π a uniformizer.

(a) The residue class field $\mathcal{O}_\mathfrak{p}/\mathfrak{M}_\mathfrak{p}$ is isomorphic to $\mathcal{O}/\mathfrak{p}$.
(b) Let S be a complete set of representatives containing 0 for the residue class field, then every $x \in L_\mathfrak{p}$ can be written uniquely in the form

$$x = \sum_{j=N}^{\infty} a_j \pi^j$$

where $N \in \mathbb{Z}$ and $a_j \in S$ for all j.
(c) In the notation of (b)

$$|x|_\mathfrak{p} = q^{-N}, \quad q \text{ is the order of the residue class field,} \quad \operatorname{ord}_\mathfrak{p}(x) = N.$$

Proposition 1.64. *The following hold:*

(a) $\mathcal{O}_\mathfrak{p}$ is compact.
(b) $L_\mathfrak{p}$ is locally compact.

Proof. For (a), we first observe that $\mathcal{O}_\mathfrak{p} = \{x \in L_\mathfrak{p} : |x|_\mathfrak{p} \leq 1\}$, hence it is closed, and therefore, is a complete metric space. Moreover, $\mathfrak{M}_\mathfrak{p} = \{x \in L_\mathfrak{p} : |x|_\mathfrak{p} < 1\}$ is open and so is every coset $a + \mathfrak{M}_\mathfrak{p}$ for any $a \in \mathcal{O}_\mathfrak{p}$. Let S be a complete set of representatives for $\mathcal{O}_\mathfrak{p}/\mathfrak{M}_\mathfrak{p}$, then S is finite and

$$\mathcal{O}_\mathfrak{p} = \bigcup_{a \in S} \left(a + \mathfrak{M}_\mathfrak{p} \right).$$

This says that $\mathcal{O}_\mathfrak{p}$ is a union of a finite number of open balls of radius 1, therefore it is totally bounded. Being a complete, totally bounded, metric space, it is compact.

(b) is now clear since for every $x \in L_\mathfrak{p}$ the open ball $B(x, 1)$ is contained in $x + \mathcal{O}_\mathfrak{p}$ which is compact.

The field $L_\mathfrak{p}$ is a finite extension of the field $\mathbb{Q}_p$. Moreover, the constructions in this example extend to the case of a general Dedekind domain, but the residue class field may not be finite, in which case the completion is not locally compact.

Example 1.65 (The Locally Compact Non-Archimedean Fields (Part 2)). Let $\mathbb{F}_q$ be a finite field with q elements where q is a power of a prime p. We consider the field of Laurent series $\mathbb{F}_q((T))$ in the indeterminate T and with coefficients in $\mathbb{F}_q$.

We already know from Proposition 1.56 that $\mathbb{F}_q((T))$ is a complete non-archimedean field. The valuation on $\mathbb{F}_q((T))$ is as follows

$$f(T) \in \mathbb{F}_q((T)), \ f(T) = \sum_{j=N}^{\infty} a_j T^j, \ ,N \in \mathbb{Z} \ a_j \in \mathbb{F}_q, \ a_N \neq 0, \ \mathrm{ord}(f(T)) = N, \ |f(T)| = q^{-N};$$

(a) The valuation ring is $\{f(T) \in \mathbb{F}_q((T)) : \mathrm{ord}(f(T)) \geq 0\} = \{f(T) \in \mathbb{F}_q((T)) : |(f(T))| \leq 1\} = \mathbb{F}_q[[T]]$, the ring of formal power series with coefficients in $\mathbb{F}_q$;
(b) The residue class field is isomorphic to $\mathbb{F}_q$.

Proposition 1.66. *The following hold:*

(a) $\mathbb{F}_q[[T]]$ *is compact in* $\mathbb{F}_q((T))$;
(b) $\mathbb{F}_q((T))$ *is locally compact.*

As in he number field case $\mathbb{F}_q[[T]]$ is closed in $\mathbb{F}_q((T))$ hence it is complete. Moreover, as the residue class field is finite, $\mathbb{F}_q[[T]]$ is contained in a finite union of open balls of radius 1, hence it is totally bounded, therefore it is compact. This proves (a). (b) follows immediately.

It should be noted that a non-archimedean valued field is locally compact if and only if (a) it is complete, (b) it is discrete and (c) its residue class field is finite. For the proof see Cassels [10].

Example 1.67 (The Field of Meromorphic Functions on a Compact Riemann Surface). Let $\mathcal{C}$ be a compact Riemann surface and $\mathcal{M}$ the field of meromorphic functions on it.

Let P be a point on $\mathcal{C}$ and $f \in \mathcal{M}$. Using a chart (U, ϕ) in a neighborhood of P, the function $f \circ \phi^{-1}$ is a meromorphic function on $\mathbb{C}$ and admits a Laurent series expansion of the form

$$\sum_{j=n_P}^{\infty} a_j z^j, \quad n_P \in \mathbb{Z}$$

which is well defined, in other words, independent of the chosen chart. The integer n_P is the order of f at P. If $n_P > 0$, f has a zero of order n_P at P, if $n_P < 0$, f has a pole of order n_P at P, and if $n_P = 0$, then f has neither a zero nor a pole at P (see [38]). The similarity with the number field case is evident.

From the theory of compact Riemann surfaces, (see, e.g., [38]), we know that non-constant meromorphic functions exist on $\mathcal{C}$. Every such a function has a finite number of zeros and poles, and more precisely, counting multiplicities (the $n_P's$), the number of zeros is the same as the number of poles.

Again, as in the number field case, it is clear how we can use a point P on $\mathcal{C}$ and the order of a function mentioned above to define a valuation on $\mathcal{M}$. We define a map $\mathrm{ord}_P : \mathcal{M} \to \mathbb{Z} \cup \{\infty\}$ as follows:

$$\mathrm{ord}_P(f) = n_P, \ \mathrm{iff} \neq 0, \quad \mathrm{ord}(0) = \infty.$$

Keeping the similarity with the number field case, the proof of the following proposition is straightforward.

Proposition 1.68. *The following hold:*

(a) ord_P *is an order function on* $\mathcal{M}$.
(b) *Let c be a real number with* $0 < c < 1$, *then* $|\cdot|_P : \mathcal{M} \to \mathbb{R}_+^*,\ f \mapsto c^{\text{ord}_P(f)}$ *is a discrete valuation.*
(c) *The valuation ring is* $\mathcal{O}_P = \{f \in \mathcal{M} : \text{ord}_P(f) \geq 0\}$, *which the set of functions that have no pole at P.*
(d) *The maximal ideal is* $\mathcal{M}_P = \{f \in \mathcal{M} : \text{ord}_P(f) > 0\}$, *which is the set of functions that vanish at P.*
(e) *A uniformizer is any function* π *whose order at P is 1.*
(f) *The residue class field* $\mathcal{O}_P / \mathcal{M}_P$ *is isomorphic to* $\mathbb{C}$.

With respect to the discrete valuation $|\cdot|_P$, $\mathcal{M}$ is a complete non-archimedean valued field with an algebraically closed residue field. We mention here that because of the correspondence between *Compact Riemann surfaces* and *Non-singular complex algebraic curves*, a discrete valuation associated to a point on a non-singular complex algebraic curve exists also, producing a complete non-archimedean valued field [38].

Example 1.69 (Non-Discrete Case: $\overline{\mathbb{Q}_p}$, *the Algebraic Closure of* $\mathbb{Q}_p$). Let $\mathbb{Q}_p$ be the field of *p*-adic numbers. It is complete with respect to the valuation $|\cdot|_p$ coming from the *p*-adic valuation on $\mathbb{Q}$ also denoted $|\cdot|_p$. Let $\overline{\mathbb{Q}_p}$ be the algebraic closure of $\mathbb{Q}_p$, then $|\cdot|_p$ can be extended to $\overline{\mathbb{Q}_p}$ (see Schikhof [46]).

Let $\alpha \in \overline{\mathbb{Q}_p}$, then α is algebraic of degree, say n over $\mathbb{Q}_p$. Let $\alpha^{(i)}, 1 \leq i \leq n$ be the conjugates of α over $\mathbb{Q}_p$. Define

$$\left|\alpha\right|_p = \left(\left|\prod_{i=1}^{n} \alpha^{(i)}\right|_p\right)^{1/n}.$$

We use the same notation $|\cdot|_p$ again on $\overline{\mathbb{Q}_p}$. This will not lead to confusion and will be justified in the next proposition. The definition makes sense as

$$\left|\prod_{i=1}^{n} \alpha^{(i)}\right|_p \in \mathbb{Q}_p.$$

For the proof of the following proposition, we refer to Schikhof [46], however, this will be partially taken up again when we deal with the non-spherical completeness of $\mathbb{C}_p$, the completion of the algebraic closure of $\mathbb{Q}_p$.

Proposition 1.70. *The following hold:*

(a) $|\cdot|_p$ *on* $\overline{\mathbb{Q}_p}$ *is a non-archimedean valuation extending* $|\cdot|_p$ *on* $\mathbb{Q}_p$.

(b) *The residue class field of $|\cdot|_p$ on $\overline{\mathbb{Q}_p}$ is $\overline{\mathbb{F}_p}$, the algebraic closure of the finite field $\mathbb{F}_p$.*

(c) *The value group of $|\cdot|_p$ on $\overline{\mathbb{Q}_p}$ is $G = \{p^{m/n} : m \in \mathbb{Z}, m \in \mathbb{N}, m \neq 0\}$. G is dense in R_+^* hence $|\cdot|_p$ is not a discrete valuation on $\overline{\mathbb{Q}_p}$.*

Remark 1.71. The following should be noted:

(a) The extension of $|\cdot|_p$ to $\overline{\mathbb{Q}_p}$ is unique.
(b) One way to obtain a non-discrete non-archimedean valued field, is to start with a complete non-archimedean valued field and use its algebraic closure as we did with $\mathbb{Q}_p$.
(c) The completion of $\overline{\mathbb{Q}_p}$ under $|\cdot|_p$ is denoted $\mathbb{C}_p$. Not only $\mathbb{C}_p$ is complete, but it is also algebraically closed, and is a non-archimedean valued field which is excellent for Analysis, in the same standing as $\mathbb{C}$, the field of complex numbers. $\mathbb{C}_p$ and $\mathbb{C}$ are isomorphic as fields but as valued fields, they are quite different. Moreover, $\mathbb{C}_p$ is separable, of infinite dimension as a vector space over $\mathbb{Q}_p$, but it is not locally compact (see Schikhof [46]).

Example 1.72 (Non-Discrete Case: The Field of Formal Puiseux Series over a field F). We begin with $F[[T]]$ the ring of formal power series, in the variable T and with coefficients in F. As in previous examples, it is a local ring with maximal ideal (T) and group of units the set of power series with non-zero constant term. The field of fractions of $F[[T]]$ is the field of formal Laurent series $F((T))$. An element $f(T)$ of $F((T))$ is of the form

$$f(T) = \sum_{j=k}^{\infty} a_j T^j, \ k \in \mathbb{Z}$$

where $a_j \in F$.

Definition 1.73. A formal Puiseux series is a series of the form $g(T^{1/n})$ in the indeterminate $T^{1/n}$ where $n \in \mathbb{N}$ and $g(T) \in F((T))$

The series contains fractional powers of T but these fractions have a common denominator. The exponents occur in increasing order and there exists a term with non-zero coefficient and with the smallest exponent. More explicitly, a Puiseux series is a series of the form

$$\sum_{j=k}^{\infty} a_j T^{j/n}$$

where $n \in \mathbb{N}$ is fixed, $k \in \mathbb{Z}$, $a_j \in F$ and $a_k \neq 0$. It is a Laurent series in the variable $T^{1/n}$.

Definition 1.74. We define the *order* of a Puiseux series

$$f(T) = \sum_{j=k}^{\infty} a_j T^{j/n}$$

with $a_k \neq 0$ to be

$$\text{ord}(f(T)) = \frac{k}{n}.$$

The value of the order function is a fraction. It is the smallest exponent occurring in the series expansion for $f(T)$.

Now by considering all possible values of n we obtain the field of formal Puiseux series.

Proposition 1.75. *For each $n \in \mathbb{N}$, let $F((T^{1/n}))$ be the field of Laurent series in the variable $T^{1/n}$ with coefficients in F and*

$$F\{\{T\}\} = \bigcup_{n=1}^{\infty} F((T^{1/n})).$$

Then $F\{\{T\}\}$ is a field, called the field of formal Puiseux series.

Proposition 1.75 is easy to prove as for each n, $F((T^{1/n}))$ is a field and for any m, $F((T^{1/n})) \subset F((T^{1/mn}))$.

Proposition 1.76. *The following hold:*

(a) The function ord : $F\{\{T\}\} \rightarrow \mathbb{Q} \cup \{\infty\}$ *defined by*

$$\text{ord}\left(\sum_{j=k}^{\infty} a_j T^j / n \right) = \frac{k}{n}, \quad \text{ord}(0) = \infty$$

is a surjective order function.

(b) Let c be a real number with $0 < c < 1$, then $|\cdot| : F\{\{T\}\} \rightarrow \mathbb{R}_+^*$ defined by

$$\left| \sum_{j=k}^{\infty} a_j T^{j/n} \right| = c^{-k/n}, \quad |0| = 0$$

is a non discrete valuation.

Proof. (a) Suppose $f(T), g(T) \in F\{\{T\}\}$ with $\text{ord}(f(T)) = k/n$ and $\text{ord}(g(T)) = l/m$, say. Then in the series for $f(T)g(T)$ the smallest exponent is $k/n + l/m$ and hence $\text{ord}(f(T)g(T)) = \text{ord}(f(T)) + \text{ord}(g(T))$. For the sum, the series will begin with an exponent which is at least the smaller between k/n and l/m and hence $\text{ord}(f(T) + g(T)) \geq \min\{\text{ord}(f(T)), \text{ord}(g(T))\}$. It is clear that the function is surjective. (b) The fact that $|\cdot|$ is a valuation follows easily from (a). The value group is the set $\{c^{r/s} : r/s \in \mathbb{Q}\}$ which is dense in $\mathbb{R}_+^*$.

1.3 Additional Properties of Non-Archimedean Valued Fields

Here we discuss some metric properties of non-archimedean valued fields. Let $\mathbb{K}$ denote a field endowed with a non-archimedean valuation, denoted $|\cdot|$. The metric d on $\mathbb{K}$ is given by

$$x, y \in \mathbb{K}, \quad d(x, y) = \left| x - y \right|.$$

Recall that for $a \in \mathbb{K}$ and for $r \in \mathbb{R}_+^*$ the open (resp. closed) ball centered at a and with radius r is denoted $B(a, r)$ (resp. $\overline{B(a, r)}$).

Lemma 1.77. *For any $a \in \mathbb{K}$ and for any $r > 0$,*

$$B(a, r) = a + B(0, r) \quad \text{and} \quad \overline{B(a, r)} = a + \overline{B(0, r)}.$$

Proof. For the first equality, let $x \in B(a, r)$, then $|x - a| < r$. Put $y = x - a$ so that $x = a + y$ and $|y| < r$. Therefore $x \in a + B(0, r)$. Next let $x \in a + B(0, r)$ then $x = a + y$ where $|y| < r$ and $x - a = y$, hence $|x - a| < r$ and $x \in B(a, r)$. The second equality is proved in a similar fashion.

Proposition 1.78. *Every ball is both open and closed.*

Proof. By Lemma 1.77, it is enough to prove the statement for a ball of arbitrary radius but centered at 0. For $r > 0$, consider the open ball $B(0, r)$. We want to show that it is closed by showing that its complement in $\mathbb{K}$ is open.

Let $x \notin B(0, r)$, so that $|x| \geq r$. We claim that the open ball $B(x, |x|)$ is contained in the complement of $B(0, r)$ in $\mathbb{K}$.

Let $z \in B(x, |x|)$ then $|x - z| < |x|$. Suppose $|z| < r \leq |x|$, then $|x - z| = |x|$ but $|x - z| < |x|$ and we get a contradiction, hence $|z| \geq r$ and therefore $z \in \mathbb{K} - B(0, r)$, and the open ball $B(x, |x|)$ is contained in the complement of $B(0, r)$. We conclude that the complement is open and $B(0, r)$ is closed. Hence $B(0, r)$ is both open and closed.

For the closed ball $\overline{B(0, r)}$, let x be an element in it so that $|x| \leq r$. Let $y \in B(x, r)$ so that $|x - y| < r$. Then $|y| = |(x - y) - x| \leq \max(|x - y|, |x|)$.

(1) If $|x| < r$, then $|y| < r$ and $B(x, r) \subset B(0, r) \subset \overline{B(0, r)}$.
(2) If $|x| = r$, then $|y| = r$ and $B(x, r) \subset \overline{B(0, r)}$.

We conclude that $\overline{B(0, r)}$ is open. Hence it is both closed and open.

In the next proposition we observe that a ball may have infinitely many radii.

Proposition 1.79. *Suppose that the valuation on $\mathbb{K}$ is discrete with order function* ord $: \mathbb{K} \to \mathbb{Z} \cup \{\infty\}$. *Let c be a real number with $0 < c < 1$ and such that for any $x \in \mathbb{K}^*$, $|x| = c^{\mathrm{ord}(x)}$. Let r be such that $0 < r < c^{-1}$, then, for any $a \in \mathbb{K}$,*

$$\overline{B(a, 1)} = \overline{B(a, r)} = B(a, r).$$

Proof. It is enough to prove the statement for the case $a = 0$, namely,

$$\overline{B(0, 1)} = \overline{B(0, r)} = B(0, r).$$

Taking into account the hypothesis and the fact that $B(0, r)$ is both open and closed, it is enough to show that

$$B(0, r) \subset \overline{B(0, 1)}.$$

Let $x \in B(0, r)$ so that $|x| < r$, hence $c^{\mathrm{ord}(x)} < r < c^{-1}$. This implies that $\mathrm{ord}(x) > -1$, hence $\mathrm{ord}(x) = 0$ or ≥ 1.

(1) If it is 0, then $|x| = 1$ and $x \in \overline{B(0, 1)}$.
(2) If it is ≥ 1, then $|x| < 1$ and $x \in B(0, 1) \subset \overline{B(0, 1)}$.

In all cases, $B(0, r) \subset \overline{B(0, 1)}$.

Proposition 1.80. *As a metric space, $\mathbb{K}$ is totally disconnected.*

Proof. Let x, y be two distinct points in $\mathbb{K}$, then there exists $r > 0$ such that $x \in B(x, r)$ and $y \notin B(x, r)$. By Proposition 1.78, both $B(x, r)$ and $\mathbb{K} - B(x, r)$ are open, disjoint, and their union is $\mathbb{K}$. This holds for any two distinct points, therefore, $\mathbb{K}$ is totally disconnected.

The valuation on $\mathbb{K}$ can be either discrete or dense. This nature of the valuation is closely related to the nature of the metric that it induces on $\mathbb{K}$.

Definition 1.81. The metric d on $\mathbb{K}$ is called *discrete* if for any sequence $(x_n, y_n)_{n \in \mathbb{N}}$ in $\mathbb{K}^2$ such that the sequence of real numbers $\{d(x_n, y_n)_n\}$ is strictly decreasing, $\lim_{n \to \infty} d(x_n, y_n) = 0$.

Proposition 1.82. *The metric d on $\mathbb{K}$ is discrete if and only if the valuation inducing it is a discrete valuation.*

Proof. Let $\mathcal{O} = \{x \in \mathbb{K} : |x| \leq 1\}$ be the valuation ring, $M = \{x \in \mathbb{K} : |x| < 1\}$ the maximal ideal, $|M| = \{|x| : x \in M\}$. Suppose d is discrete and let $\alpha = \sup |M|$, clearly $\alpha \neq 0$. Suppose there exists a sequence $\{x_n\}_n$ in M such that the sequence $\{|x_n|\}_n$ is strictly increasing with $\lim_{n \to \infty} |x_n| = \alpha$, then taking the inverses we obtain a sequence $\{y_n\}_n$ in $\mathbb{K}$ such that the sequence $\{|y_n|\}_n$ is strictly decreasing and converges to $1/\alpha \neq 0$. This contradicts the hypothesis since $|y_n| = d(0, y_n)$. Therefore, for any sequence $\{x_n\}_n$ in M such that $\lim_{n \to \infty} |x_n| = \alpha$, the sequence $\{|x_n|\}$ is stationary, in other words, there exists an integer N such that for any $n \geq N$, $|x_n| = |x_N|$. This, now, implies that $\alpha \in |M|$ and there exists $\pi \in M$ such that $|\pi| = \alpha$.

Now for any $x \in M$, $|\frac{x}{a}| \leq 1$ therefore $x = \pi y$ for some $y \in \mathcal{O}$. We conclude that M is a principal ideal generated by π and by Proposition 1.29, the valuation is discrete.

Next suppose the valuation is discrete and let π be a uniformizer, i.e., a generator for the maximal ideal M. Let $\{(x_n, y_n)\}_n$ be a sequence in $\mathbb{K}^2$ such that the sequence $\{d(x_n, y_n)\}_n$ is strictly decreasing.

$$\text{Put } z_n = x_n - y_n \in \mathbb{K}$$

then

$$d(x_n, y_n) = |z_n|$$

and the sequence

$$\{|z_n|\}_n \text{ is strictly decreasing}$$

But $z_n = u_n \pi^{a_n}$ where u_n is a unit in the valuation ring $\mathcal{O}$, $a_n \in \mathbb{Z}$ and $|z_n| = |\pi|^{a_n}$. Since $|\pi| < 1$ and the sequence $\{|z_n|\}_n$ is strictly decreasing, the sequence of integers $\{a_n\}_n$ is strictly increasing and hence $\lim_{n \to \infty} a_n = \infty$. Again with $|\pi| < 1$, this implies that $\lim |z_n|_{n \to \infty} = 0$.

Definition 1.83. The metric d on $\mathbb{K}$ is called *dense* if for every ball B in $\mathbb{K}$, the set $\{d(x, y) : x, y \in B\}$ is dense in the closed interval $[0, d(B)]$ where $d(B)$ stands for the diameter of B.

Proposition 1.84. *The metric d on $\mathbb{K}$ is dense if and only the valuation inducing it is dense.*

Recall that the valuation is called dense if it is not discrete and this means, the value group $|\mathbb{K}^*|$ is dense in $\mathbb{R}_+^*$.

Proof. Suppose the metric is dense and that the valuation is discrete. Let c be such that $0 < c < 1$ and $|\mathbb{K}^*| = \{c^n : n \in \mathbb{Z}\}$. Consider the ball $B = B(0, 1)$, which is actually the maximal ideal M of the valuation. We make the following simple observations:

(a) $\{d(x, y) : x, y \in B\} = \{|x| : x \in B\} = |M| = \{c, c^2, c^3, \ldots\}$;
(b) $d(B) = \sup(|M|) = c$.

Now by hypothesis, the set $|M| = \{|x| : x \in B\}$ is dense in the closed interval $[0, c]$. However, the open interval $(c^2, c) \subset [0, c]$ has an empty intersection with $|M|$. This contradiction implies that the valuation is not discrete and hence it is dense.

Conversely, suppose the valuation is dense. First we consider the case of a ball of the form $B = B(0, r)$ where $r > 0$. Then

$$\{d(x, y) : x, y \in B\} = \{|x - y| : x, y \in B\} = \{|z| : z \in B\} \text{ and } [0, d(B)] \subset [0, r].$$

Let (a, b) be an open interval contained in $[0, d(B)]$. Since the valuation is dense, there exists $x \in \mathbb{K}^*$ such that $|x| \in (a, b)$. Now we have the following inequalities

$0 < a < |x| < b < d(B) \leq r$ which shows that $x \in B$. Therefore we can conclude that $\{d(x,y) : x, y \in B\}$ is dense in $[0, d(B)]$. Now moving to a ball of the form $B = B(a, r)$, we use Lemma 1.77 and write

$$B = B(a, r) = a + B(0, r).$$

Let us denote the ball $B(0, r)$ by B'. Then

$$\left\{ \left| x - y \right| : x, y \in B \right\} = \left\{ \left| t - u \right| : t, u \in B' \right\} = \left\{ \left| z \right| : z \in B' \right\}.$$

It follows that $d(B) = d(B')$. Now let (c, d) be an open interval contained in $[0, d(B)]$, then since $\{|z| : z \in B'\}$ is dense in $[0, d(B)]$, there exists $x \in B'$ such that $0 < c < |x| < d < d(B) \leq r$. Hence $\{|x - y| : x, y \in B\}$ is dense in $[0, d(B)]$.

An important aspect of the metric space structure on a non-archimedean valued field is that of *spherical completeness*.

Definition 1.85. The non-archimedean valued field $\mathbb{K}$ is called *spherically complete* if every decreasing sequence of balls in $\mathbb{K}$ has a non-empty intersection.

This notion is closely related to that of ordinary completeness. Spherical completeness implies ordinary completeness but not conversely. The following proposition gives examples of spherically complete non-archimedean valued fields.

Proposition 1.86. *The following hold:*

(a) *Suppose $\mathbb{K}$ is endowed with a non-archimedean valuation and with the induced metric, it is spherically complete, then $\mathbb{K}$ is a complete metric space.*
(b) *Let $\mathbb{K}$ be endowed with a discrete valuation and suppose that with the induced metric, it is complete, then, $\mathbb{K}$ is spherically complete.*

Proof. (a) Let $\{x_n\}_n$ be a Cauchy sequence in $\mathbb{K}$. For every positive integer k there exists an integer n_k such that $\forall n \geq n_k$, $|x_n - x_{n_k}| \leq \frac{1}{k}$. We obtain a strictly increasing sequence of integers $\{n_k\}_k$ and a subsequence $\{x_{n_k}\}_k$. Consider the sequence of closed balls $\left\{ \overline{B(x_{n_k}, 1/k)} \right\}_k$. This sequence is decreasing. Indeed, let $x \in \overline{B(x_{n_{k+1}}, 1/(k+1))}$, then $|x_{n_{k+1}} - x| \leq \frac{1}{k+1}$. We observe that $|x_{n_{k+1}} - x_{n_k}| \leq \frac{1}{k}$. Therefore

$$\left| x_{n_k} - x \right| \leq \max \left\{ \left| x_{n_k} - x_{n_{k+1}} \right|, \left| x_{n_{k+1}} - x \right| \right\} \leq \frac{1}{k}.$$

Hence $x \in \overline{B(x_{n_k}, 1/k)}$ and

$$\overline{B(x_{n_{k+1}}, 1/(k+1))} \subset \overline{B(x_{n_k}, 1/k)}.$$

Since, by hypothesis, $\mathbb{K}$ is spherically complete, there exists $a \in \mathbb{K}$ such that

$$a \in \bigcap_{k=1}^{\infty} \overline{B(x_{n_k}, 1/k)}.$$

Then it is clear that $\lim_{k \to \infty} x_{n_k} = a$.

Lemma 1.87. $\lim_{n \to \infty} x_n = a$.

Proof. Let $\epsilon > 0$ be given. (1) There exists k_1 such that for all $k \geq k_1$, $|x_{n_k} - a| < \epsilon$. (2) There exists k_2 such that $\frac{1}{k_2} < \epsilon$ and for all $n \geq n_{k_2}, |x_n - x_{n_{k_2}}| < \frac{1}{k_2} < \epsilon$. Let $k_3 = \max\{k_1, k_2\}$ and $N = n_{k_3}$, then

$$\forall n \geq N, \quad \left| x_n - a \right| \leq \max \left\{ \left| x_n - x_N \right|, \left| x_N - a \right| \right\} < \epsilon$$

and we conclude that $\mathbb{K}$ is a complete metric space.

(3) We suppose that the valuation is discrete and that $\mathbb{K}$ is complete. Let π be a uniformizer so that $|x| = |\pi|^{\mathrm{ord}(x)}$ for any $x \in \mathbb{K}^*$.

Lemma 1.88. *Let B be a ball in $\mathbb{K}$ then, the diameter $d(B) = \sup\{d(x, y) : x, y \in B\}$, is attained, in other words, there exists $(x_1, y_1) \in B^2$ such that $d(x_1, y_1) = d(B)$.*

Proof. Let r be the radius of B, a its center. Let $n = \lfloor \log_{|\pi|}(r) \rfloor$, in other words n is an integer such that

$$n \leq \log_{|\pi|}(r) < n + 1.$$

Then for any $x, y \in B$, $a - x = u\pi^k$, $a - y = v\pi^l$ where u, v are units, of absolute value 1, and k, l are integers. Now, $|a - x| \leq r, a - y| \leq r$ and by the definition of n and the fact that $|\pi| < 1$, we find that

$$k \geq n \quad \text{and} \quad l \geq n$$

hence,

$$d(x, y) = \left| x - y \right| \leq \max \left\{ \left| \pi \right|^k, \left| \pi \right|^l \right\} \leq \left| \pi \right|^n.$$

Therefore $d(B) \leq |\pi|^n$. But if we take $x_1 = a + \pi^n$, $y_1 = a$ then

$$x_1, y_1 \in B \quad \text{and} \quad d(x_1, y_1) = \left| \pi \right|^n = d(B).$$

Next let $B_1 \supset B_2 \ldots \supset B_k \supset \ldots$ be a decreasing sequence of balls in $\mathbb{K}$. We may assume that the sequence is strictly decreasing, in other words $B_k \neq$

B_{k+1} for all k. From Lemma 1.88, and for each k, there exists $(x_k, y_k) \in B_k^2$ such that $d(x_k, y_k) = d(B_k)$ and the sequence $\{d(x_k, y_k)\}_k$ is strictly decreasing. Since the valuation is discrete, by Proposition 1.82 the induced metric d is discrete and therefore $\lim_{k \to \infty} d(x_k, y_k) = 0$. Now, for any $\epsilon > 0$, there exists an integer N such that for all $k \geq N$, $d(B_k) < \epsilon$. For integers $j \geq k \geq N$, we have

$$d(B_j) < \epsilon, \quad d(B_k) < \epsilon \text{ and both } x_j, x_k \text{ lie in } B_k.$$

Therefore, $d(x_j, x_k) < \epsilon$ and the sequence $\{x_k\}_k$ is a Cauchy sequence. Since $\mathbb{K}$ is complete the sequence converges to some $x \in \mathbb{K}$. Clearly $x \in \bigcap_{k=1}^{\infty}$ and therefore $\bigcap_{k=1}^{\infty} \neq \emptyset$ and $\mathbb{K}$ is spherically complete.

A standard example of a non-archimedean field which is *complete* but not *spherically complete* is $\mathbb{C}_p$, the completion of the algebraic closure of $\mathbb{Q}_p$. In order to show this, we prove a result which is interesting in its own right and can be generalized. We refer to Artin [2] for more.

Proposition 1.89. *Let p be a prime and $\mathbb{C}_p$ the completion of the algebraic closure of $\mathbb{Q}_p$, then $\overline{\mathbb{Q}}$, the algebraic closure of $\mathbb{Q}$ is dense in $\mathbb{C}_p$.*

Proof. Let $B = B(a, r)$ be a ball centered at a and of radius $r > 0$ in $\mathbb{C}_p$. Since the algebraic closure of $\mathbb{Q}_p$ is dense in $\mathbb{C}_p$, there exists $\alpha \in B$ which is algebraic over $\mathbb{Q}_p$ and we can assume to be non-zero. Let

$$f(x) = x^n + a_{n-1}x^{n-1} + \ldots + a_1 x + a_0 \in \mathbb{Q}_p[x]$$

be the minimal polynomial of α over $\mathbb{Q}_p$.

Let A be a real number satisfying $A \geq \max\{1, |a_0|, |a_1|, \ldots, |a_{n-1}|\}$.

Lemma 1.90. $|\alpha| \leq A$.

Proof. Suppose, to the contrary, that $|\alpha| > A$, then for any $k = 0, \ldots, n-1$, $|\alpha^n| > |a_k \alpha^k|$ therefore $0 = |\alpha^n + a_{n-1}\alpha^{n-1} + \ldots + a_1\alpha + a_0| = |\alpha|^n$. This is a contradiction, hence the Lemma is proven.

Let ϵ be a positive real number satisfying $\epsilon < r^n$. Then, since $\mathbb{Q}$ is dense in $\mathbb{Q}_p$, there exist $b_0, \ldots, b_{n-1} \in \mathbb{Q}$ such that

$$\text{for any } k, \quad |a_k - b_k| < \frac{\epsilon}{A^n}.$$

Consider the polynomial

$$g(x) = x^n + b_{n-1}x^{n-1} + \ldots + b_1 x + b_0 \in \mathbb{Q}[x] = (x - \beta_1)(x - \beta_2) \ldots (x - \beta_n)$$

where $\beta_1, \ldots \beta_n$ are the roots of $g(x)$. We observe:

$$|g(\alpha) - f(\alpha)| = \left| \sum_{k=1}^{n-1} (b_k - a_k)\alpha^k \right|$$

$$\leq \max \left\{ \left| b_k - a_k \right| \left| \alpha \right|^k : k = 1, \ldots, n-1 \right\}$$

$$< \frac{\epsilon}{A^n} . A^n$$

$$< \epsilon.$$

But since $f(\alpha) = 0$, we find

$$\left| g(\alpha) \right| = \left| \alpha - \beta_1 \right| . \left| \alpha - \beta_2 \right| \ldots \left| \alpha - \beta_n \right| < \epsilon$$

which implies that there exists $i, 0 \leq i \leq n-1$, such that $|\alpha - \beta_i| < \epsilon^{1/n} < r$ and hence $|a - \beta_i| < r$. Since $\beta_i \in \overline{\mathbb{Q}}$, $B = B(a, r)$ is an arbitrary ball in $\mathbb{C}_p$, we see that $\overline{\mathbb{Q}}$ is dense in $\mathbb{C}_p$.

The following theorem, due to Krull, is very important. Although we do not use it in the later parts of the book, it is quoted here to illustrate, even in a small context, the extension of valuations (see Schikhof [46]).

Theorem 1.91. *(a) (existence) Let $\mathbb{K}$ be a non-archimedean valued field and $\mathbb{L}$ an extension of $\mathbb{K}$, then, there exists a valuation on $\mathbb{L}$ which extends that of $\mathbb{K}$.*

(b) (uniqueness) Let $\mathbb{K}$ be a complete non-archimedean valued field and $\mathbb{L}$ an algebraic extension of $\mathbb{K}$, then, there exists a unique valuation on $\mathbb{L}$ which extends that of $\mathbb{K}$.

Let $| \cdot |$ be the p-adic valuation on $\mathbb{Q}_p$ for a prime p and let $\overline{\mathbb{Q}_p}$ be the algebraic closure of $\mathbb{Q}_p$. Every $x \in \overline{\mathbb{Q}_p}$ has a minimal polynomial $f(T) \in \mathbb{Q}_p[T]$, say

$$f(T) = T^n + a_{n-1}T^{n-1} + \ldots + a_1 T + a_0$$

where $n \in \mathbb{N}$ and $a_i \in \mathbb{Q}_p$ for $i = 0, \ldots, n-1$, and if we call $x = x_1, x_2, \ldots, x_n$ the roots of $f(T)$ then

$$f(T) = T^n + a_{n-1}T^{n-1} + \ldots + a_1 T + a_0 = (T - x_1)(T - x_2) \ldots (T - x_n).$$

Proposition 1.92. *In the context above, and with an abuse of notation, let $| \cdot | : \overline{\mathbb{Q}_p} \to \mathbb{R}_+^*$ be defined by $|x| = |a_0|^{1/n}$. Then, $| \cdot |$ is the unique non-archimedean valuation on $\overline{\mathbb{Q}_p}$ that extends the valuation on $\mathbb{Q}_p$. Moreover, $|\cdot|$ is a dense valuation.*

Proof. First, we observe that by Krull's uniqueness theorem, if a valuation on $\overline{\mathbb{Q}_p}$ extends that of $\mathbb{Q}_p$, then it is unique because $\mathbb{Q}_p$ is complete and $\overline{\mathbb{Q}_p}$ is algebraic over $\mathbb{Q}_p$.

We now verify that $|\cdot|$ is indeed a non-archimedean valuation. Parts (1) and (2) of Definition 1.1 are easily verified. What remains to verify is (3). Let $x \in \overline{\mathbb{Q}_p}$ be as above and suppose $|x| \leq 1$. We observe that in the notation above $|x| = |x_1| = |x_2| = \ldots = |x_n| = |a_0|^{1/n} \leq 1$. As a consequence, since the a_i's are the symmetric functions of $x_1, x_2, \ldots, x_n$, we find that $|a_i| \leq 1$, $i = 0, 2, \ldots, n-1$. Now consider $y = x + 1$. Then the minimal polynomial of y over $\mathbb{Q}_p$ is $f(T - 1)$ and hence

$$\left| y \right| = f(-1)$$

$$= \left| (-1)^n + a_n - 1(-1^{n-1} + \ldots + a_1(-1) + a_0 \right|$$

$$\leq \max \left\{ \left| a_i \right| : i = 0, 1, \ldots, n-1 \right\}$$

$$\leq 1.$$

This proves Part (3) of Definition 1.1 and $|\cdot|$ is a non-archimedean valuation on $\overline{\mathbb{Q}_p}$. It is clear that it extends the valuation on $\mathbb{Q}_p$.

We now prove that this valuation is dense.

Proposition 1.93. $|(\overline{\mathbb{Q}_p})^*| = \{p^s : s \in \mathbb{Q}\}$ *and hence the valuation* $|\cdot|$ *is dense.*

Proof. Let $x \in (\overline{\mathbb{Q}_p})^*$ and let $f(T) = T^n + a_{n-1}T^{n-1} + \ldots + a_1 T + a_0$ be the minimal polynomial of x over $\mathbb{Q}_p$. Then $|x| = |a_0|^{1/n}$. Since $a_0 \in \mathbb{Q}_p$, $|a_0| = p^m$, $m \in \mathbb{Z}$. Therefore $|x| = p^{m/n}$.

Next consider $p^{m/n}$, $m, n \in \mathbb{Z}$. There exists $a \in \mathbb{Q}_p^*$ such that $|a| = p^m$. Let $x \in (\overline{\mathbb{Q}_p})^*$ such that $x^n = a$, then $|x| = |a|^{1/n} = p^{m/n}$.

Proposition 1.94. *The field* $\mathbb{C}_p$ *is not spherically complete.*

Proof. The valuation on $\mathbb{C}_p$ is dense. By Proposition 1.89 , $\mathbb{C}_p$ contains a countable dense subset. Let $a_1, a_2, \ldots, a_n, \ldots$ be one such countable dense subset. Let α be a positive real number and consider a decreasing sequence of positive real numbers $r_1 > r_2 > \ldots > r_n > \ldots > \alpha$. If all balls of radius r_1 contain a_1, then $\mathbb{C}_p = B(a_1, r_1)$ which is impossible, hence, there exists a ball $B_1 = B(b_1, r_1)$ centered at some b_1 of radius r_1 such that $a_1 \notin B_1$. If all balls of radius r_2 contained in B_1 contain a_2, then, we would have $B_1 = B(b_1, r_1) = B(a_2, r_2) = B(b_1, r_2)$ however, since the valuation is dense, there exists $t \in \mathbb{K}$ such that $r_2 < |t| < r_1$. The element $x = b_1 + t \in B(b_1, r_1) \setminus B(b_1, r_2)$ this contradiction implies that there exists a ball $B_2 = B(b_2, r_2)$ centered at some $b_2 \in B_1$, of radius r_2 such that $B_1 \supset B_2$, $a_2 \notin B_2$. And so on, we obtain a decreasing sequence of balls $B_1 \supset B_2 \supset \ldots \supset B_n \supset \ldots, a_n \notin B_n$. But since $r_n > \alpha$ for all n, then for each n, B_n contains a ball of radius α. Suppose $\bigcap_{n=1}^{\infty} B_n \neq \emptyset$. Let x be in this intersection, then $B = B(x, \alpha) \subset B_n$, for every n.

But since the set $\{a_1, a_2, \ldots, a_n, \ldots\}$ is dense, there exists k such that $a_k \in B$ and hence $a_k \in B_k$, which is a contradiction. Therefore $\bigcap_{n=1}^{\infty} B_n = \emptyset$, and $\mathbb{C}_p$ is not spherically complete. The above argument can be generalized to any separable non-archimedean metric space whose metric is induced by a dense valuation (see Schikhof [46]).

1.4 Some Remarks on Krull Valuations

Let $(\Gamma, +, \leq)$ be a totally ordered abelian group. This means that

(i) $(\Gamma, +)$ is an abelian group;
(ii) for any $\alpha, \beta \in \Gamma$, either $\alpha \leq \beta$ or $\beta \leq \alpha$;
(iii) for any $\alpha, \beta, \gamma \in \Gamma$, $\alpha \leq \beta$ implies $\alpha + \gamma \leq \beta + \gamma$.

We point out the following Proposition:

Proposition 1.95. *If Γ is a totally ordered abelian group, then Γ is torsion-free.*

Proof. Without loss of generality let $\gamma > 0$ be in Γ. Suppose, by induction, that $n\gamma > 0$, then $n\gamma + \gamma > \gamma > 0$, hence $(n+1)\gamma > 0$. We may conclude that for any integer $n \geq 1$, $n\gamma > 0$, therefore Γ is torsion-free.

Definition 1.96. Let $\mathbb{K}$ be a field. A mapping $v : \mathbb{K} \to \Gamma \cup \{\infty\}$ is called a *Krull valuation* if it satisfies the following:

(a) $v(x) = \infty$ if and only if $x = 0$;
(b) $v(xy) = v(x) + v(y)$ for any $x, y \in \mathbb{K}^*$;
(c) $v(x + y) \geq \min\{v(x), v(y)\}$ for any $x, y \in \mathbb{K}$.

The symbol ∞ satisfies the following:

(i) $\infty \notin \Gamma$;
(ii) for any $\gamma \in \Gamma$, $\gamma + \infty = \infty + \gamma = \infty$;
(iii) for any $\gamma \in \Gamma$, $\gamma < \infty$.

The following Proposition is immediate

Proposition 1.97. *The Krull valuation satisfies:*

(a) $v(-x) = v(x)$ *for any* $x \in \mathbb{K}$;
(b) $v(x^{-1}) = -v(x)$ *for any* $x \in \mathbb{K}^*$;
(c) $v(x + y) = \min\{v(x), v(y)\}$ *if* $v(x) \neq v(y)$.

Proof. (a) $v(1) = v(1) + v(1)$ hence $v(1) = 0$. Moreover, $0 = v(1) = v(-1) + v(-1)$ hence $v(-1) = 0$. Now $v(-x) = v(-1) + v(x) = v(x)$.
(b) $0 = v(1) = v(xx^{-1}) = v(x) + v(x^{-1})$ hence $v(x^{-1}) = -v(x)$.
(c) Suppose $v(x) < v(y)$ so that $v(x + y) \geq v(x)$. Then $v(x + y - y) = v(x) \geq \min\{v(x + y), v(y)\}$ but since $v(x) < v(y)$ we see that $v(x + y) < v(y)$ and hence $v(x) \geq v(x + y)$. We conclude that $v(x + y) = v(x)$.

Definition 1.98. The *valuation ring* of the Krull valuation v on $\mathbb{K}$ is

$$A_v = \left\{ x \in \mathbb{K} : v(x) \geq 0 \right\}.$$

Proposition 1.99. *The valuation ring A_v is a local ring with unique maximal ideal*

$$M_v = \left\{ x \in \mathbb{K} : v(x) > 0 \right\}.$$

Proof. Let I be an ideal in A_v. Either all elements of I have positive valuations, in which case, $I \subset M_v$ or there exists a non-zero element of I with zero valuation, in which case, by Proposition 1.97, its inverse in $\mathbb{K}$ also has zero valuation and therefore lies in A_v, and this implies that $I = A_v$.

The residue field is A_v/M_v, the group of units is $U_v = \{x \in \mathbb{K} : v(x) = 0\} = A_v \backslash M_v$ and the value group is $v(\mathbb{K}^*)$.

The valuation v induces a topology on the field $\mathbb{K}$ by considering

$$\left\{ O_\varepsilon : \varepsilon \in \Gamma \right\}$$

as a neighborhood base of $0 \in \mathbb{K}$, where

$$O_\varepsilon = \left\{ x \in \mathbb{K} : v(x) > \varepsilon \right\}.$$

Consequently, a sequence $(x_j)_{j \in \mathbb{N}} \subset \mathbb{K}$ converges to 0 for this topology if only if

$$v(x_j) \to \infty \ \ as \ j \to \infty.$$

Example 1.100. We construct an example of a Krull valuation whose value group is not contained in $\mathbb{R}_+^*$. Let $\mathbb{Q}[T, S]$ be the ring of polynomials in two variables T and S with coefficients in $\mathbb{Q}$. Every polynomial $f(T, S)$ can be written as

$$f(T, S) = T^a (B_0(S) + B_1(S)T + \ldots + B_n(S)T^n)$$

where $a, n \in \mathbb{N}$ and $B_i(S) \in \mathbb{Q}[S]$ for $i = 0, 1, \ldots, n$.
   Now $B_0(S)$ can be written as

$$B_0(S) = S^b (u_0 + u_1 S + \ldots + u_m S^m)$$

where $b, m \in \mathbb{N}$ and $u_j \in \mathbb{Q}$ for $j = 0, 1, \ldots, m$. Therefore $f(T, S)$ can be written as

$$f(T, S) = T^a (S^b (u_0 + u_1 S + \ldots + u_m S^m) + B_1(S)T + \ldots + B_n(S)T^n).$$

We put $\Gamma = \mathbb{Z} \times \mathbb{Z}$ and we define a total order on Γ using the lexicographic order

$$(a,b) \leq (c,d) \quad \text{if and only if} \quad (a < c) \quad \text{or} \quad (a = c \text{ and } b < d).$$

Next we define

$$v : \mathbb{Q}[T,S] \to \Gamma \cup \{\infty\}$$

by

$$v(0) = \infty, \quad v(f(T,S)) = (a,b).$$

Using the same notation, we extend v to $\mathbb{Q}(T,S)$, the field of rational functions in T and S with coefficients in $\mathbb{Q}$, by putting

$$v\left(\frac{f(T,S)}{g(T,S)}\right) = v(f(T,S)) - v(g(T,S)).$$

As always the symbol ∞ satisfies the usual properties.

Proposition 1.101. v *is a Krull valuation.*

Proof. First $v(0) = \infty$ and if $\frac{f(T,S)}{g(T,S)} \neq 0$ then $v\left(\frac{f(T,S)}{g(T,S)}\right) \neq \infty$. Next

$$v\left(\frac{f(T,S)}{g(T,S)} \cdot \frac{h(T,S)}{k(T,S)}\right) = v\left(f(T,S)h(T,S)\right) - v\left(g(T,S)k(T,S)\right)$$

$$= \left(v\left(f(T,S)\right) - v\left(g(T,S)\right)\right) + \left(v\left(h(T,S)\right) - v\left(k(T,S)\right)\right)$$

$$= v\left(\frac{f(T,S)}{g(T,S)}\right) + v\left(\frac{h(T,S)}{k(T,S)}\right).$$

Lemma 1.102. *For polynomials $f(T,S)$ and $g(T,S)$,*

$$v\left(f(T,S) + g(T,S)\right) \geq \min\left\{v(f(T,S)), v(g(T,S))\right\}.$$

Proof. Let $f(T,S) = T^a[S^b(u_0 + \ldots + u_m S^m) + B_1(S)T + \ldots + B_n(S)T^n]$ and $g(T,S) = T^c[S^d(w_0 + \ldots + w_k S^k) + C_1(S)T + \ldots + C_l T^l]$ so that $v(f(T,S)) = (a,b)$ and $v(g(T,S)) = (c,d)$. Suppose that $(a,b) \leq (c,d)$.

Case 1: $a < c$ then

$$f(T,S) + g(T,S) = T^a(S^b(u_0 + \ldots + u_m S^m) + B_1(S)T + \ldots + B_n(S)T^n)$$
$$+ T^{c-a}(S^d(w_0 + \ldots + w_k S^k) + C_1(S)T + \ldots + C_l T^l)$$

and we find that

$$v\Big(f(T,S)+g(T,S)\Big)=(a,b)=\min\Big\{v\Big(f(T,S),v(g(T,S))\Big)\Big\}.$$

Case 2: $a=c$ and $b<d$ then

$$f(T,S)+g(T,S)=T^a(S^b(u_0+\ldots+u_m S^m+S^{d-b}(w_o+\ldots+w_k S^k))+B_1(S)T$$
$$+\ldots+B_n(S)T^n+C_1(S)T+\ldots+C_l(S)T^l)$$

and we find that

$$v\Big(f(T,S)+g(T,S)\Big)=(a,b)=\min\Big\{v\Big(f(T,S),v(g(T,S))\Big)\Big\}.$$

Using Lemma 1.102, we easily find that

$$v\left(\frac{f(T,S)}{g(T,S)}+\frac{h(T,S)}{k(T,S)}\right)\geq\min\left\{v\left(\frac{f(T,S)}{g(T,S)}\right),v\left(\frac{h(T,S)}{k(T,S)}\right)\right\}.$$

1.5   Bibliographical Notes

The basic material for this chapter and some examples come from the following sources: Artin [2], Cassels [10], Schikhof [46], Miranda [38], Attimu [4], Attimu and Diagana [6], Endler [23], and Schilling [47].

Chapter 2
Non-Archimedean Banach Spaces

In this chapter we gather some basic facts about non-archimedean Banach spaces, with a special emphasis on the so-called p-adic Hilbert space. Again the results here are well-known and will serve as background for the operator theory developed in later chapters.

Let $\mathbb{K}$ denote a *complete* non-archimedean valued field. The valuation on $\mathbb{K}$ will be denoted $|\cdot|$.

2.1 Non-Archimedean Norms

In this section we introduce and study basic properties of non-archimedean norms and non-archimedean normed spaces.

Definition 2.1. Let $\mathbb{E}$ be a vector space over $\mathbb{K}$. A *non-archimedean norm* on $\mathbb{E}$ is a map $\|\cdot\| : \mathbb{E} \to \mathbb{R}_+^*$ satisfying

(1) $\|x\| = 0$ if and only if $x = 0$;
(2) $\|\lambda x\| = |\lambda| \|x\|$ for any $x \in \mathbb{E}$ and any $\lambda \in \mathbb{K}$;
(3) $\|x + y\| \leq \max\{\|x\|, \|y\|\}$ for any $x, y \in \mathbb{E}$.

Property (3) of Definition 2.1 is referred to as the *ultrametric* or *strong triangle inequality*.

Definition 2.2. A non-archimedean normed space is a pair $(\mathbb{E}, \|\cdot\|)$ where $\mathbb{E}$ is a vector space over $\mathbb{K}$ and $\|\cdot\|$ is a non-archimedean norm on $\mathbb{E}$.

Unless there is an explicit mention of the contrary, all norms that are considered in this chapter are non-archimedean. Here are some first examples of non-archimedean norms on some vector spaces.

© The Author(s) 2016
T. Diagana, F. Ramaroson, *Non-Archimedean Operator Theory*,
SpringerBriefs in Mathematics, DOI 10.1007/978-3-319-27323-5_2

Example 2.3. The valuation on $\mathbb{K}$ itself is a non-archimedean norm.

Example 2.4. Consider the Cartesian product $\mathbb{K}^n$ with $n \in \mathbb{N}$ and define

$$\left\| (x_1, \ldots, x_n) \right\| = \max \left\{ \left| x_i \right| : 1 \leq i \leq n \right\}.$$

Then this is a non-archimedean norm on $\mathbb{K}^n$.

Example 2.5. For each $i \in \mathbb{N}$ let $\mathbb{K}^{(i)} = \mathbb{K}$ and consider

$$P = \prod_{i=0}^{\infty} \mathbb{K}^{(i)}.$$

Then P is the direct product of a countable copies of $\mathbb{K}$. The set P is naturally a vector space over $\mathbb{K}$. Note that an element of P is just a sequence of elements of $\mathbb{K}$ of the form $(x_i)_{i \in \mathbb{N}}$.

Let

$$L^{\infty}(\mathbb{K}) = \left\{ (x_i)_{i \in \mathbb{N}} \in P : \ (x_i)_{i \in \mathbb{N}} \text{ is bounded} \right\}.$$

Then $L^{\infty}(\mathbb{K})$ is a subspace of P. Define

$$\left\| (x_i)_i \right\|_{\infty} = \sup \left\{ \left| x_i \right| : i \in \mathbb{N} \right\},$$

then, this is a non-archimedean norm on $L^{\infty}(\mathbb{K})$.

Example 2.6. With the notations of Example 2.5, consider

$$S = \sum_{i=0}^{\infty} \mathbb{K}^{(i)},$$

then S is the direct sum of a countable copies of $\mathbb{K}$ and it is a subspace of P. Note that an element of S is a sequence of the form $(x_i)_{i \in \mathbb{N}}$ such that $x_i \in \mathbb{K}$, $x_i = 0$ for almost all (for all except for a finite number) $i \in \mathbb{N}$.

Define

$$\left\| (x_i)_{i \in \mathbb{N}} \right\| = \max \left\{ \left\| x_i \right\| : i \in \mathbb{N} \right\},$$

then, this is a well-defined non-archimedean norm on S.

Example 2.7. Let $\mathbb{X}$ be a set and let $B(\mathbb{X}, \mathbb{K})$ be the set all *bounded* functions on $\mathbb{X}$ with values in $\mathbb{K}$, then with the operations

$$(f + g)(x) = f(x) + g(x), \ f, g \in B(\mathbb{X}, \mathbb{K}), \ x \in \mathbb{X}$$
$$(\lambda f)(x) = \lambda f(x), \ \lambda \in \mathbb{K}, \ f \in B(\mathbb{X}, \mathbb{K}), \ x \in \mathbb{X}$$

Clearly, $B(\mathbb{X}, \mathbb{K})$ is a vector space over $\mathbb{K}$.

Define the sup-norm

$$\|f\|_\infty := \sup \left\{ |f(x)| : x \in \mathbb{K} \right\},$$

then this is a non-archimedean norm on $B(\mathbb{X}, \mathbb{K})$.

Example 2.8. Let $(\mathbb{E}_i, \| \cdot \|_i)$, $i = 1, 2$ be non-archimedean normed spaces over $\mathbb{K}$. Let $L(\mathbb{E}_1, \mathbb{E}_2)$ be the set of all $\mathbb{K}$-linear maps $A : \mathbb{E}_1 \to \mathbb{E}_2$, then it is naturally a vector space over $\mathbb{K}$. Let $B(\mathbb{E}_1, \mathbb{E}_2)$ be the set all $\mathbb{K}$-linear maps $A : \mathbb{E}_1 \to \mathbb{E}_2$ satisfying the following: there exists $C \geq 0$ such that for all $x \in \mathbb{E}_1$,

$$\|Ax\|_2 \leq C \|x\|_1.$$

Then $B(\mathbb{E}_1, \mathbb{E}_2)$ is a subspace of $L(\mathbb{E}_1, \mathbb{E}_2)$. Define for all $A \in B(\mathbb{E}_1, \mathbb{E}_2)$,

$$\|A\| = \sup_{x \in \mathbb{E}_1 \setminus \{0\}} \frac{\|A(x)\|_2}{\|x\|_1}.$$

Then, this is a non-archimedean norm on $B(\mathbb{E}_1, \mathbb{E}_2)$.

Proposition 2.9. *Let $(\mathbb{E}, \| \cdot \|)$ be a non-archimedean normed space. For $x, y \in \mathbb{E}$,*

$$\|x + y\| = \max \left\{ \|x\|, \|y\| \right\}, \ \text{if } \|x\| \neq \|y\|.$$

Proof. Suppose that $\|x\| < \|y\|$ so that $\max\{\|x\|, \|y\|\} = \|y\|$, then by Definition 1.1, $\|x + y\| \leq \|y\|$. Now

$$\|y\| = \|x + y - x\| \leq \max \left\{ \|x + y\|, \|x\| \right\}.$$

But since $\|y\| > \|x\|$, we must have

$$\max \left\{ \|x + y\|, \|x\| \right\} = \|x + y\|,$$

and therefore

$$\|y\| \leq \|x + y\|$$

and the conclusion follows.

Definition 2.10. Let $(\mathbb{E}, \|\cdot\|)$ be non-archimedean normed space and S be a non-empty subset of E. The set S is said to be *bounded* if the set of real numbers $\{\|x\| : x \in S\}$ is bounded.

Definition 2.11. A sequence $(x_i)_{i\in\mathbb{N}}$ in the normed space $(\mathbb{E}, \|\cdot\|)$ *converges* (strongly) to $x \in \mathbb{E}$ and we write

$$\lim_{i\to\infty} x_i = x$$

if the sequence of real numbers $(\|x_i - x\|)_{i\in\mathbb{N}}$ converges to 0.

Definition 2.12. A series $\sum_{i=0}^{\infty} x_i$ in $(\mathbb{E}, \|\cdot\|)$ *converges* to $x \in \mathbb{E}$ and we write

$$\sum_{i=0}^{\infty} x_i = x$$

if the sequence of partial sums $(s_n)_{n\in\mathbb{N}}$

$$s_n = \sum_{i=0}^{n} x_i, \quad n \in \mathbb{N}$$

converges to x.

Proposition 2.13. *Let $(\mathbb{E}, \|\cdot\|)$ be a non-archimedean normed space over $\mathbb{K}$. If the sequence $(x_i)_{i\in\mathbb{N}}$ converges in $\mathbb{E}$, then it is bounded.*

Proof. Suppose $(x_i)_{i\in\mathbb{N}}$ converges to x, then the sequence of real numbers $(\|x_i - x\|)_i$ converges in $\mathbb{R}$, therefore is bounded. It follows that the set $\{x_i : i \in \mathbb{N}\}$ is bounded as a subset of $\mathbb{E}$.

2.2 Non-Archimedean Banach Spaces

In this section we introduce non-archimedean Banach spaces, discuss their properties and illustrate with examples.

Let $(\mathbb{E}, \|\cdot\|)$ be a non-archimedean normed space, then a metric d can be defined on $\mathbb{E}$ to give it the topology of a metric space. This metric is defined by

$$x, y \in \mathbb{E}, \; d(x, y) := \left\| x - y \right\|.$$

Proposition 2.14. *The strong triangle inequality translates as follows:*

$$\text{for } x, y, z \in \mathbb{E}, \; d(x, y) \le \max\left\{ d(x, z), d(y, z) \right\}.$$

Definition 2.15. A normed space $(\mathbb{E}, \|\cdot\|)$ is called a *Banach space* if it is complete with respect to the natural metric induced by the norm

$$d(x, y) = \|x - y\|, \quad x, y \in \mathbb{E}.$$

The spaces $\mathbb{K}$, $\mathbb{K}^n$, $\sum_{i=0}^{\infty} \mathbb{K}^{(i)}$, $L^{\infty}(\mathbb{K})$, $B(\mathbb{X}, \mathbb{K})$, $B(\mathbb{E}_1, \mathbb{E}_2)$ with their respective norms are Banach spaces.

Proposition 2.16. *(1) a close subspace of a Banach space is a Banach space;* *(2) the direct sum of two Banach spaces is a Banach space.*

Proof. (1) is clear. (2) the norm on the direct sum is defined by $\|(x, y)\| = \max\{\|x\|, \|y\|\}$. From there, the proof is also clear.

We next define the norm on the quotient space using the same notation for both the space and its quotient.

Definition 2.17. Let $\mathbb{E}$ be a Banach space and $\mathbb{V}$ a closed subspace of $\mathbb{E}$. Let $P : \mathbb{E} \to \mathbb{E}/\mathbb{V}$ be the quotient map. Define

$$\|Px\| = d(x, \mathbb{V}), \quad x \in \mathbb{E},$$

where

$$d(x, B) = \inf \left\{ d(x, z) : z \in \mathbb{V} \right\} = \inf \left\{ \|x - z\| : z \in \mathbb{V} \right\}$$

is the distance from x to $\mathbb{V}$.

Remark 2.18. This norm is well defined because $Px = Py$ if and only if $x - y \in \mathbb{V}$, moreover $\|Px\| \leq \|x\|$ for any $x \in \mathbb{E}$.

Proposition 2.19. *The norm in Definition 2.17 is a non-archimedean norm on $\mathbb{E} \backslash \mathbb{V}$.*

Proof. (1) First $\|0\| = \|P(0)\| = 0$ since $0 \in \mathbb{V}$. Next, if $\|Px\| = 0$ then $d(x, \mathbb{V}) = 0$ hence $x \in \mathbb{V}$, and $Px = 0$.
(2) For any $\lambda \in \mathbb{K}^*$,

$$\begin{aligned}
\|\lambda Px\| &= \|P(\lambda x)\| \\
&= \inf \left\{ \|\lambda x - z\| : z \in \mathbb{V} \right\} \\
&= |\lambda| \inf \left\{ \left\| x - \frac{z}{\lambda} \right\| : z \in \mathbb{V} \right\} \\
&= |\lambda| \inf \left\{ \|x - y\| : y \in V \right\} \\
&= |\lambda| \|Px\|.
\end{aligned}$$

(3) For $x, y \in \mathbb{E}$, since $\mathbb{V}$ is closed, there exist $z_1, z_2, z_3 \in \mathbb{V}$ such that

$$\left\| Px \right\| = \left\| x - z_1 \right\|, \quad \left\| Py \right\| = \left\| y - z_2 \right\|, \quad \left\| P(x+y) \right\| = \left\| x + y - z_3 \right\|$$

$$\begin{aligned}
\left\| P(x) + P(y) \right\| &= \left\| P(x+y) \right\| \\
&= \left\| x + y - z_3 \right\| \\
&\leq \left\| (x+y) - (z_1 + z_2) \right\| \quad \text{(because } (z_1 + z_2) \in \mathbb{V}) \\
&= \left\| (x - z_1) + (y - z_2) \right\| \\
&\leq \max\left\{ \left\| x - z_1 \right\|, \left\| y - z_2 \right\| \right\} \\
&= \max\left\{ \left\| Px \right\|, \left\| Py \right\| \right\}.
\end{aligned}$$

There will be more examples in the later parts of the book.

An example that plays a very important role in the theory of non-archimedean Banach space, is the following:

Example 2.20. Let $c_0(\mathbb{K})$ denote the set of all sequences $(x_i)_{i \in \mathbb{N}}$ in $\mathbb{K}$ such that

$$\lim_{i \to \infty} \left| x_i \right| = 0.$$

Then, $c_0(\mathbb{K})$ is a vector space over $\mathbb{K}$ and

$$\left\| (x_i)_{i \in \mathbb{N}} \right\| = \sup_{i \in \mathbb{N}} \left| x_i \right|$$

is a non-archimedean norm for which $(c_0(\mathbb{K}), \| \cdot \|)$ is a Banach space.

Another important example which will play a central role in the book is now defined. It is a modified version of $c_0(\mathbb{K})$.

Example 2.21. Let $\omega = (\omega_i)_{i \in \mathbb{N}}$ be a sequence of *non-zero* elements in $\mathbb{K}$. We define the space $\mathbb{E}_\omega$, by

$$\mathbb{E}_\omega = \left\{ x = (x_i)_{i \in \mathbb{N}} : \forall i, x_i \in \mathbb{K} \text{ and } \lim_{i \to \infty} \left(\left| \omega_i \right|^{1/2} \left| x_i \right| \right) = 0 \right\}.$$

On $\mathbb{E}_\omega$, we define

$$x = (x_i)_{i \in \mathbb{N}} \in \mathbb{E}_\omega, \quad \left\| x \right\| = \sup_{i \in \mathbb{N}} \left(\left| \omega_i \right|^{1/2} \left| x_i \right| \right).$$

Then, $(\mathbb{E}_\omega, \|\cdot\|)$ is a non-archimedean Banach space.

Definition 2.22. The Banach space $\mathbb{E}_\omega$ of Example 2.21, equipped with its norm, is called a *p-adic Hilbert space*.

The space $\mathbb{E}_\omega$ will play a central role in the book and will be the subject of further studies in this chapter and in later chapters.

Example 2.23. In reference to Example 2.8, if we take $\mathbb{E}_1 = \mathbb{E}_2 = \mathbb{E}$, the non-archimedean normed space $B(\mathbb{E}, \mathbb{E})$ is denoted $B(\mathbb{E})$ and consists of all $\mathbb{K}$-linear maps $A : \mathbb{E} \to \mathbb{E}$ (also called "*linear operators*") satisfying

$$\exists C \geq 0 \text{ such that } \forall x \in \mathbb{E}, \ \left\|Ax\right\| \leq C\left\|x\right\|.$$

Recall that the norm on $B(\mathbb{E})$ is

$$\left\|A\right\| = \sup_{x \in E \backslash \{0\}} \frac{\left\|Ax\right\|}{\left\|x\right\|}.$$

Then $B(\mathbb{E})$ is also a very important non-archimedean Banach space and will be thoroughly discussed in Chap. 3. It is called the space of *bounded* or *continuous* linear operators on $\mathbb{E}$.

Example 2.24. Again in reference to Example 2.8, if we take $\mathbb{E}_1 = \mathbb{E}$ a Banach space over $\mathbb{K}$ and $\mathbb{E}_2 = \mathbb{K}$, then the Banach space $B(\mathbb{E}, \mathbb{K})$ is called the *dual* of $\mathbb{E}$ and denoted $\mathbb{E}^*$. The dual $\mathbb{E}^*$, then, is the space of bounded linear functionals on $\mathbb{E}$. If $\xi \in \mathbb{E}^*$, then

$$\left\|\xi\right\|_* = \sup_{x \in \mathbb{E} \backslash \{0\}} \frac{\left|\langle \xi, x\rangle\right|}{\left\|x\right\|}.$$

With this norm $\mathbb{E}^*$ is a Banach space.

Let $\mathbb{E}^{**}$ be the dual of $\mathbb{E}^*$, then, there is a natural $\mathbb{K}$-linear map:

$$j_{\mathbb{E}} : \mathbb{E} \to \mathbb{E}^{**} \text{ such that } \forall x \in \mathbb{E}, \ j_{\mathbb{E}}(x)(\xi) = \langle \xi, x\rangle, \ \forall \xi \in \mathbb{E}^*.$$

We now consider some properties of Banach spaces that will be useful later.

Proposition 2.25. *Let* $(\mathbb{E}, \|\cdot\|)$ *be a Banach space. The series* $\sum_{i=0}^{\infty} x_i$ *converges in* $\mathbb{E}$ *if and only if the sequence of general terms* $(x_i)_{i \in \mathbb{N}}$ *converges to* 0.

Proof. Suppose that the series converges, then it is clear that the general term converges to 0. Conversely, suppose that

$$\lim_{i \to \infty} x_i = 0.$$

This means that for any $\varepsilon > 0$ there exists N such that for $i > N$, $\|x_i\| < \varepsilon$.

Consider the sequence of partial sums $(s_k)_{k\in\mathbb{N}}$ where

$$s_k = \sum_{i=0}^{k} x_i.$$

Then for $n > m > N$

$$\|s_n - s_m\| = \|x_{m+1} + \ldots + x_n\|$$
$$\leq \max\left\{\|x_j\| : m+1 \leq j \leq n\right\}$$
$$< \varepsilon.$$

The sequence of partial sums $(s_k)_{k\in\mathbb{N}}$ is a Cauchy sequence in the Banach space $\mathbb{E}$ hence it converges.

As in the classical case, we have the following definition.

Definition 2.26. Let $\mathbb{E}$ be a vector space over $\mathbb{K}$ and $\|\cdot\|_1$ and $\|\cdot\|_2$ two non-archimedean norms on $\mathbb{E}$ for each of which $\mathbb{E}$ is a Banach space. The two norms are said to be *equivalent* if there exist positive constants c_1 and c_2 such that for any $x \in \mathbb{E}$,

$$c_1\|x\|_1 \leq \|x\|_2 \leq c_2\|x\|_1.$$

Proposition 2.27. *On a finite dimensional Banach space over $\mathbb{K}$, all non-archimedean norms are equivalent.*

Proof. We use induction on the dimension n. If $n = 1$, let $\|x\|_0 = |x|$ be the norm determined by the absolute value. Now let $\|\cdot\|$ be any norm on $\mathbb{K}$, then for any $x \in \mathbb{K}$,

$$\|x\| = |x|\|1\| = c\|x\|_0, \quad \text{with } c = \|1\|$$

which implies that $\|\cdot\|$ is equivalent to $\|\cdot\|_0$. Suppose that the proposition is true for a space of dimension $(n-1)$. Let $\mathbb{E}$ be of dimension n and let $\{e_1, \ldots, e_n\}$ be a basis for $\mathbb{E}$. First we have the natural norm on $\mathbb{E}$ which is

$$x \in \mathbb{E}, \quad x = \sum_{i=1}^{n} x_i e_i, \quad \|x\|_0 = \max\left\{|x_i| : 1 \leq i \leq n\right\}.$$

Let $\|\cdot\|$ be any norm on $\mathbb{E}$. We want to show that $\|\cdot\|$ is equivalent to $\|\cdot\|_0$.

For any $x = \sum_{i=1}^{n} x_i e_i$ we have

$$\left\| x \right\| = \left\| \sum_{i=1}^{n} x_i e_i \right\| \leq \max \left\{ \left| x_i \right| \left\| e_i \right\| : 1 \leq i \leq n \right\} \leq C \left\| x \right\|_0,$$

where $C = \max\{ \|e_i\| : 1 \leq i \leq n \}$ and we find

$$\left\| x \right\| \leq C \left\| x \right\|_0.$$

To obtain the other inequality which will complete the equivalence, we let $\mathbb{V}$ be the subspace of $\mathbb{E}$ generated by $\{e_1, \ldots, e_{n-1}\}$, then

$$x = y + x_n e_n$$

where $y = \sum_{i=1}^{n-1} x_i e_i \in \mathbb{V}$. We note that $\mathbb{V}$ is a closed subspace of $\mathbb{E}$, being the set of all vectors in E whose n-th component is zero. Therefore, it follows that

$$a = \inf \left\{ \left\| z + e_n \right\| : z \in \mathbb{V} \right\} > 0$$

then

$$\left\| x_n^{-1} y + e_n \right\| \geq a > 0.$$

Put

$$b = a \left\| e_n \right\|^{-1} \text{ so that } b \leq 1.$$

Suppose first that $x_n \neq 0$, then

$$\left\| e_n \right\|^{-1} \left\| x_n^{-1} y + e_n \right\| \geq b.$$

Now

$$\left\| x \right\| = \left| x_n \right| \left\| e_n \right\| \left(\left\| e_n \right\|^{-1} \left\| x_n^{-1} y + e_n \right\| \right) \geq b \left\| x_n e_n \right\|$$

and we find

$$\left\| x \right\| \geq b \left\| x_n e_n \right\|.$$

Lemma 2.28. $\|x\| \geq b\|y\|$.

Proof. Suppose that $\|x\| < b\|y\|$ hence $\|y + x_n e_n\| < b\|y\|$ and since $b \leq 1$ we find that

$$\|y + x_n e_n\| < \|y\|$$

which implies that

$$\|x_n e_n\| = \|(y + x_n e_n) - y\| = \|y\|$$

and since $\|y + x_n e_n\| \geq b\|x_n e_n\|$ we get a contradiction.

Now we have

$$\|x\| \geq b|x_n|\|e_n\| \quad \text{and} \quad \|x\| \geq b\|y\|.$$

By induction, there exist constants b' and b'' such that

$$\|x\| \geq bb'|x_n| \quad \text{and} \quad \|x\| \geq bb'' \max\left\{|x_i| : 1 \leq i \leq (n-1)\right\}.$$

Let $C = \min\{bb', bb''\}$. Then,

$$\|x\| \geq C \max\left\{|x_i| : 1 \leq i \leq n\right\} = C\|x\|_0.$$

Suppose next that $x_n = 0$. In this case, we still have

$$\|x\| \geq b\|y\|$$

and the same argument carries on, hence, $\|\cdot\|$ is equivalent to $\|\cdot\|_0$.

2.3 Free Banach Spaces

In this section we define and discuss properties of Banach spaces which have bases. Let $\mathbb{E}$ be a Banach space over $\mathbb{K}$.

Definition 2.29. A family $(v_i)_{i \in I}$ of vectors in $\mathbb{E}$ indexed by a set I converges to 0 and we write

$$\lim_{i \in I} v_i = 0$$

if

$$\forall \varepsilon > 0, \left\{ i \in I : \left\| v_i \right\| \geq \varepsilon \right\} \text{ is finite.}$$

Definition 2.30. Let $v \in E$ and let $(v_i)_{i \in I}$ be a family of elements of $\mathbb{E}$ indexed by the set I. We say that v is the sum of the family $(v_i)_{i \in I}$ and we write

$$\sum_{i \in I} v_i = v$$

if $\forall \varepsilon > 0$, there exists a finite subset $J_0 \subset I$ such that for any finite $J \subset I$, $J \supseteq J_0$

$$\left\| \sum_{i \in J} v_i - v \right\| \leq \varepsilon.$$

In this situation, we also say that the family $(v_i)_{i \in I}$ is summable and its sum is v.

Proposition 2.31. *Let the family $(v_i)_{i \in I}$ be summable in $\mathbb{E}$ with sum $v \in \mathbb{E}$, then*

$$\lim_{i \in I} v_i = 0.$$

Proof. Given $\varepsilon > 0$, let $H = \{i \in I : \|v_i\| \geq \varepsilon\}$. Since the family $(v_i)_{i \in I}$ is summable with sum v, there exists a finite subset J_0 of I such that for any finite subset J of I containing J_0,

$$\left\| \sum_{i \in J} v_i - v \right\| \leq \varepsilon.$$

Let $j \in I \setminus J_0$ and consider $J = J_0 \cup \{j\}$ then

$$\left\| \sum_{i \in J} v_i - v \right\| \leq \varepsilon.$$

Since

$$\left\| \sum_{i \in J_0} v_i - v \right\| \leq \varepsilon$$

it follows that

$$\max \left\{ \left\| \sum_{i \in J} v_i - v \right\|, \left\| \sum_{i \in J_0} v_i - v \right\| \right\} \leq \varepsilon$$

which implies that

$$\left\| v_j \right\| \le \varepsilon.$$

Since this holds for any $j \notin J_0$, we conclude that

$$H \subset J_0$$

hence H is finite and therefore $\lim_{i \in I} v_i = 0$.

Definition 2.32. A *basis* for $\mathbb{E}$ is a family of elements of $\mathbb{E}$, $\{e_i : i \in I\}$ indexed by a set I such that for every $x \in \mathbb{E}$ there exists a unique family $(x_i)_{i \in I}$ of elements in $\mathbb{K}$ such that

$$\sum_{i \in I} x_i e_i = x.$$

In this situation, in view of Proposition 2.31, $\lim_{i \in I} x_i e_i = 0$.

Example 2.33. In Example 2.20 we introduced the Banach space $c_0(\mathbb{K})$. It has the following basis $\{e_i : i \in \mathbb{N}\}$, where $e_0 = (1, 0, 0, \ldots), e_1 = (0, 1, 0, 0, \ldots) \ldots$ in other words, e_i is the sequence all whose terms are 0 except the i-th term which is equal to 1. If $x = (x_i)_{i \in \mathbb{N}} \in c_0(\mathbb{K})$ then

$$x = \sum_{i=0}^{\infty} x_i e_i$$

and

$$\lim_{i \to \infty} \left| x_i \right| = 0.$$

Let p be a prime, and suppose $\mathbb{K}$ is a finite extension of $\mathbb{Q}_p$, the field of p-adic numbers. We let $\mathbb{Z}_p$ be the ring of p-adic integers.

Definition 2.34. For each $i \in \mathbb{N}$ we define the *Mahler function M_i* to be the function

$$M_i : \mathbb{Z}_p \to \mathbb{K}, \quad M_0(x) = 1 \text{ and for } i > 0, \, M_i(x) = \binom{x}{i} = \frac{x(x-1)\ldots(x-i+1)}{i!}, \quad x \in \mathbb{Z}_p.$$

The function M_i satisfies the following:

(1) $M_i(j) = 0$ if j is an integer with $j < n$;
(2) $M_i(i) = 1$;
(3) $M_i(x)$ is a polynomial function of degree i.

Let $C(\mathbb{Z}_p, \mathbb{K})$ be the $\mathbb{K}$-vector space of continuous functions from the compact set $\mathbb{Z}_p$ to $\mathbb{K}$, equipped with the sup-norm

$$\left\| f \right\|_\infty := \sup_{z \in \mathbb{Z}_p} \left| f(z) \right|.$$

Example 2.35. The space $C(\mathbb{Z}_p, \mathbb{K})$ is a free Banach space because of the following classic theorem of Mahler for whose proof we refer to [46] or [53].

Theorem 2.36. *The following hold:*

(1) For each $i \in \mathbb{N}$, $M_i \in C(\mathbb{Z}_p, \mathbb{K})$ and $\|M_i\| = 1$;
(2) For each $f \in C(\mathbb{Z}_p, \mathbb{K})$ there exists a unique sequence $(a_i)_{i \in \mathbb{N}} \subset \mathbb{K}$ such that

$$f(x) = \sum_{i=0}^{\infty} a_i M_i(x), \quad x \in \mathbb{Z}_p.$$

The series converges uniformly and

$$\left\| f \right\|_\infty = \max \left\{ \left| a_i \right| : i \in \mathbb{N} \right\};$$

(3) If $(a_i)_i \in \mathbb{N} \in c_0(\mathbb{K})$ then, the function

$$f(x) = \sum_{i=0}^{\infty} a_i M_i, \quad x \in \mathbb{Z}_p$$

defines an element of $C(\mathbb{Z}_p, \mathbb{K})$.

We now introduce the notion of orthogonality:

Definition 2.37. We say that $x, y \in \mathbb{E}$ are orthogonal to each other if

$$\left\| ax + by \right\| = \max \left\{ \left\| ax \right\|, \left\| by \right\| \right\}, \quad \text{for any } a, b \in \mathbb{K}.$$

This definition is clearly symmetric and generalizes as follows:

Definition 2.38. Let $(v_i)_{i \in I}$ be a family of vectors in $\mathbb{E}$. We say that the family is *orthogonal* if for any $J \subset I$ and for any family $(a_i)_{i \in J}$ of elements of $\mathbb{K}$ such that $\lim_{i \in J} a_i v_i = 0$,

$$\left\| \sum_{i \in J} a_i x_i \right\| = \max \left\{ \left\| a_i x_i \right\| : i \in J \right\}.$$

Definition 2.39. An *orthogonal basis* for the Banach space $\mathbb{E}$ is a base which is an orthogonal family.

This means, then, that a family $\{e_i : i \in I\}$ is an orthogonal basis if and only if

(1) For every $x \in E$, there exists a unique family $(x_i)_{i \in I} \subset \mathbb{K}$ such that $x = \sum_{i \in I} x_i e_i$;
(2) $\left\| x \right\| = \max \left\{ \left\| x_i e_i \right\| : i \in I \right\}$;

The orthogonal basis $\{e_i : i \in I\}$ is called an *orthonormal* basis if $\|e_i\| = 1$ for all $i \in I$.

Remark 2.40. The space $c_0(\mathbb{K})$ has a natural orthonormal base, namely, $e_i : i = 0, 1, \ldots$ where the sequence $e_i = (\delta_{i,j}) \in \mathbb{N}$ and $\delta_{i,j}$ is the Kronecker symbol.

Remark 2.41. The sequence of Mahler functions $\{M_i : i = 0, 1, \ldots\}$ forms an orthonormal basis of $C(\mathbb{Z}_p, \mathbb{K})$.

2.4 The *p*-adic Hilbert Space $\mathbb{E}_\omega$

In this section we discuss properties of the so-called *p*-adic Hilbert space $\mathbb{E}_\omega$ which will be the focus of operator theory in the later chapters. We follow Diarra [20] closely.
Recall from Example 2.21 and Definition 2.22 that given $\omega = (\omega_i)_{i \in \mathbb{N}} \subset \mathbb{K}^*$,

$$\mathbb{E}_\omega = \left\{ x = (x_i)_{i \in \mathbb{N}} : x_i \in \mathbb{K}, \ \forall i \in \mathbb{N}, \ \lim_{i \to \infty} \left| \omega_i \right|^{1/2} \left| x_i \right| = 0 \right\}.$$

The space $\mathbb{E}_\omega$ is equipped with the norm

$$x = (x_i)_{i \in \mathbb{N}} \in \mathbb{E}_\omega, \quad \left\| x \right\| := \sup_{i \in \mathbb{N}} \left| \omega_i \right|^{1/2} \left| x_i \right|.$$

Another characterization of $\mathbb{E}_\omega$ is the following:

Proposition 2.42. $x = (x_i)_{i \in \mathbb{N}} \in \mathbb{E}_\omega$ *if and only if* $\lim_{i \to \infty} x_i^2 \omega_i = 0$.

Proposition 2.43. *The normed space* $(\mathbb{E}_\omega, \| \cdot \|)$ *is a free Banach space with orthogonal basis* $\{e_i : i = 0, 1, \ldots\}$ *where* $e_i = (\delta_{i,j})_{j \in \mathbb{N}}$ *and* $\delta_{i,j}$ *is the Kronecker symbol. For each* $i \in \mathbb{N}$

$$\left\| e_i \right\| = \left| \omega_i \right|^{1/2}.$$

Remark 2.44. The orthogonal basis $\{e_i : i = 0, 1, 2, \ldots\}$ is called the *canonical basis* of $\mathbb{E}_\omega$.

Proposition 2.45. *Let* $\langle \cdot, \cdot \rangle : \mathbb{E}_\omega \times \mathbb{E}_\omega \to \mathbb{K}$ *be defined as follows: for* $x = (x_i)_{i\in\mathbb{N}}$, $y = (y_i)_{i\in\mathbb{N}}$

$$\langle x, y \rangle = \sum_{i=0}^{\infty} x_i y_i \omega_i.$$

Then

(1) $\langle x, y \rangle$ *is well-defined, i.e., the series converges in* $\mathbb{K}$;
(2) $\langle \cdot, \cdot \rangle$ *is symmetric, bilinear form on* $\mathbb{E}_\omega$;
(3) $\langle \cdot, \cdot \rangle$ *satisfies the Cauchy–Schwarz inequality, namely,*

$$\left| \langle x, y \rangle \right| \leq \left\| x \right\| \cdot \left\| y \right\|;$$

(4) $\langle \cdot, \cdot \rangle$ *is continuous.*

Proof. (2) is clear and (4) follows from (3).
(1)

$$\lim_{i\to\infty} \left| x_i \right| \left| y_i \right| \left| \omega_i \right| = \lim_{i\to\infty} \left| x_i \right| \left| \omega_i \right|^{1/2} \left| y_i \right| \left| \omega_i \right|^{1/2}$$

$$= \left(\lim_{i\to\infty} \left| x_i \right| \left| \omega_i \right|^{1/2} \right) \cdot \left(\lim_{i\to\infty} \left| y_i \right| \left| \omega_i \right|^{1/2} \right)$$

$$= 0 \ \text{ since } x, y \in \mathbb{E}_\omega.$$

(3)

$$\left| \langle x, y \rangle \right| = \left| \sum_{i=0}^{\infty} x_i y_i \omega_i \right|$$

$$\leq \sup_{i\in\mathbb{N}} \left| x_i \right| \left| \omega_i \right|^{1/2} \left| y_i \right| \left| \omega_i \right|^{1/2}$$

$$\leq \sup_{i\in\mathbb{N}} \left| x_i \right| \left| \omega_i \right|^{1/2} \cdot \sup_{i\in\mathbb{N}} \left| y_i \right| \left| \omega_i \right|^{1/2}$$

$$= \left\| x \right\| \cdot \left\| y \right\|.$$

Moreover, we have:

Proposition 2.46. *The following hold:*

(1) $\langle \cdot, \cdot \rangle$ *is non-degenerate;*

(2) $\langle x, x \rangle = \sum_{i=0}^{\infty} x_i^2 \omega_i$;

(3) $\langle e_i, e_j \rangle = \delta_{i,j} \omega_i$ for $i, j \in \mathbb{N}$;

(4) $\langle x, e_k \rangle = x_k \omega_k$.

Proof. We prove only (1). Suppose $\langle x, y \rangle = 0$ for all $y \in \mathbb{E}_\omega$. Then, in particular, for any k, if $y = e_k$, we find that $x_k \omega_k = 0$ which implies that $x_k = 0$ as $\omega_k \neq 0$. Since this holds for all k, $x = 0$.

The space $\mathbb{E}_\omega$ endowed with the above-mentioned norm and inner product, is called a *p-adic* (or *non-archimedean*) *Hilbert space*. In contrast with classical Hilbert spaces, the norm on $\mathbb{E}_\omega$ does not stem from the inner product. Further, the space $\mathbb{E}_\omega$ contains isotropic vectors, that is, vectors $x \in \mathbb{E}_\omega$ such that $\langle x, x \rangle = 0$ while $x \neq 0$. Let us construct one of those isotropic vectors of $\mathbb{E}_\omega$. To simply things, suppose $\mathbb{K} = \mathbb{Q}_p$ where p is a prime satisfying $p \equiv 1 \pmod{4}$ and let $\omega = (\omega_i)_{i \in \mathbb{N}}$, where $\omega_0 = 1$, $\omega_1 = 1$, $\omega_i = p^i$ for $i \geq 2$. If we consider the nonzero vector $x = (x_i)_{i \in \mathbb{N}} \in \mathbb{E}_\omega$, given by, $x_0 = 1$, $x_1 = \sqrt{-1} \in \mathbb{Q}_p$, $x_i = 0$, $i \geq 2$, it easily follows that $\langle x, x \rangle = 0$ while $\|x\| = 1$.

In the case of the *p-adic* Hilbert space $\mathbb{E}_\omega$, the definition of an orthogonal basis becomes the following:

Definition 2.47. $\{h_i : i = 0, 1, \ldots\} \subset \mathbb{E}_\omega$ is an orthogonal basis for $\mathbb{E}_\omega$ if

(1) For every $x \in \mathbb{E}_\omega$ there exists a unique sequence $(x_i)_{i \in \mathbb{N}}$ such that $x = \sum_{i=0}^\infty x_i h_i$; and

(2) $\|x\| = \sup_{i \in \mathbb{N}} |x_i| \|h_i\|$.

An example of an orthogonal basis is the canonical basis $\{e_i : i = 0, 1, \ldots\}$ of Proposition 2.43.

We next consider perturbations of orthogonal bases in $\mathbb{E}_\omega$. Namely, if $\{h_i : i = 0, 1, \ldots\}$ is an orthogonal basis and if $\{f_i : i = 0, 1, \ldots\}$ is another sequence of vector of $\mathbb{E}_\omega$, not necessarily an orthogonal basis, such that the difference $f_i - h_i$ is small in a certain sense; we investigate conditions under which the family $\{f_i : i = 0, 1, \ldots\}$ is a basis of $\mathbb{E}_\omega$.

We refer to Chap. 3 for deeper results where invertible operators play crucial roles, however, we have the following:

Proposition 2.48. *Let* $\{f_i : i = 0, 1, \ldots\}$ *be an orthogonal basis in* E *and let* $\{h_j : j = 0, 1, \ldots\}$ *be a basis in* E *such that*

$$\|h_i - f_i\| < \|f_i\|$$

then $\{h_j : j = 0, 1, \ldots\}$ *is an orthogonal basis.*

Proof. Let $\{a_i : i = 0, 1, \ldots\}$ be a sequence in $\mathbb{K}$ such that $\lim_{i \to \infty} a_i h_i = 0$, and let $x = \sum_{i=0}^\infty a_i h_i$. We observe that the assumption $\|h_i - f_i\| < \|f_i\|$ implies that $\|h_i\| = \|f_i\|$ for all i. This, in turn, implies that $\lim_{i \to \infty} a_i f_i = 0$. Let $y = \sum_{i=0}^\infty a_i f_i$, then,

$$\left\|x - y\right\| = \left\|\sum_{i=0}^{\infty} a_i(h_i - f_i)\right\|$$

$$\leq \sup_{i \in \mathbb{N}} |a_i| \left\|h_i - f_i\right\|$$

$$< \sup_{i \in \mathbb{N}} |a_i| \left\|f_i\right\|$$

$$= \left\|\sum_{i \in \mathbb{N}} a_i f_i\right\|$$

$$= \left\|x\right\|,$$

and therefore, $\left\|x - y\right\| < \left\|x\right\|$.

Now this implies that $\left\|y\right\| = \left\|(y - x) + x\right\| = \left\|x\right\|$ and therefore

$$\left\|x\right\| = \sup_{i \in \mathbb{N}} |a_i| \left\|h_i\right\|$$

and hence $\{h_i : i = 0, 1, \ldots\}$ is an orthogonal basis.

We conclude this background chapter with a structure theorem for $\mathbb{E}_\omega$.

Proposition 2.49. *Let $\pi \in \mathbb{K}$ such that $|\pi| < 1$. There exists a sequence*

$$\{f_i : i = 0, 1, \ldots\} \subset \mathbb{E}_\omega$$

satisfying the following

(1) For each $i \in \mathbb{N}$, f_i is a scalar multiple of e_i such that

$$|\pi| \leq \left\|f_i\right\| \leq 1;$$

(2) $\{f_i : i = 0, 1, \ldots\}$ is an orthogonal basis for $\mathbb{E}_\omega$ in the sense of Definition 2.47.

Proof. (1) follows from the following

Lemma 2.50. *For every $x \in \mathbb{E}_\omega$, $x \neq 0$, there exists $x' \in \mathbb{E}_\omega$ a scalar multiple of x such that*

$$|\pi| \leq \left\|x\right\| \leq 1.$$

Proof. Since $|\pi| < 1$, there exists an integer n such that

$$\left|\pi\right|^{n+1} \leq \left\|x\right\| \leq \left|\pi\right|^{n}.$$

Dividing through by $\left|\pi\right|^{n}$ gives the result.

(2) From (1) we can write for each $i \in \mathbb{N}$, $f_i = \lambda_i e_i$ where $\lambda_i \in \mathbb{K}$ and $\lambda_i \neq 0$. Let $x \in E_\omega$, then, there exists $(x_i)_{i\in\mathbb{N}} \subset \mathbb{K}$ such that

$$x = \sum_{i=0}^{\infty} x_i e_i = \sum_{i=0}^{\infty} \frac{x_i}{\lambda_i} f_i$$

$$= \sum_{i=0}^{\infty} y_i f_i, \quad \text{with } y_i = \frac{x_i}{\lambda_i} \in \mathbb{K}.$$

Next suppose

$$x = \sum_{i=0}^{\infty} y_i f_i = \sum_{i=0}^{\infty} z_i f_i, \quad y_i, z_i \in \mathbb{K}, \ \forall i.$$

Then

$$x = \sum_{i=0}^{\infty} y_i \lambda_i e_i = \sum_{i=0}^{\infty} z_i \lambda_i e_i$$

which implies $\lambda_i y_i = \lambda_i z_i$ for all i, and hence $y_i = z_i$ for all i. Moreover, for $x \in E_\omega$, $x = \sum_{i=0}^{\infty} y_i f_i$

$$\left\|x\right\| = \left\|\sum_{i=0}^{\infty} y_i f_i\right\|$$

$$= \left\|\sum_{i=0}^{\infty} y_i \lambda_i e_i\right\|$$

$$= \sup_{i\in\mathbb{N}} \left|y_i\right| \left|\lambda_i\right| \left\|e_i\right\|$$

$$= \sup_{i\in\mathbb{N}} \left|y_i\right| \left\|f_i\right\| \quad \text{as } \left|\lambda_i\right| \left\|e_i\right\| = \left\|f_i\right\|$$

and hence $\{f_i : i = 0, 1, \ldots\}$ is an orthogonal basis in the sense of Definition 2.47.

Theorem 2.51. *The p-adic Hilbert space $\mathbb{E}_\omega$ is bicontinuously isomorphic to $c_0(\mathbb{K})$.*

Proof. We show that there exists a continuous, linear, bijection $\Phi : c_0(\mathbb{K}) \to \mathbb{E}_\omega$ whose inverse $\Psi : \mathbb{E}_\omega \to c_0(\mathbb{K})$ is also continuous.

We use the orthogonal basis $\{f_i : i = 0, 1, \ldots\}$ of Proposition 2.49 for $\mathbb{E}_\omega$. Let $\Phi : c_0(\mathbb{K}) \to \mathbb{E}_\omega$ be defined as follows, $x = (x_i)_{i \in \mathbb{N}} \in c_0(\mathbb{K})$, $\lim\limits_{i} \left| x_i \right| = 0$,

$$\Phi(x) = \sum_{i=0}^{\infty} x_i f_i.$$

Since $\left\| f_i \right\| \leq 1$, then

$$\lim_{i \to \infty} \left| x_i \right| \left\| f_i \right\| \leq \lim_{i \to \infty} \left| x_i \right| = 0$$

hence Φ is well-defined and is clearly linear.

$$\left\| \Phi(x) \right\| = \left\| \sum_{i=0}^{\infty} x_i f_i \right\|$$

$$= \sup_{i \in \mathbb{N}} \left| x_i \right| \left\| f_i \right\|$$

$$\leq \sup_{i \in \mathbb{N}} \left| x_i \right|, \quad \text{as } \left\| f_i \right\| \leq 1$$

$$= \left\| x \right\|$$

hence Φ is continuous.

Let $\Psi : \mathbb{E}_\omega \to c_0(\mathbb{K})$ be defined as follows, $y = \sum_{i=0}^{\infty} y_i f_i$, $\lim\limits_{i \to \infty} \left| y_i \right| \left\| f_i \right\| = 0$, and

$$\Psi(y) = (y_i)_{i \in \mathbb{N}}.$$

Since $\left\| f_i \right\| \geq \left| \pi \right|$, then

$$0 = \lim_{i \to \infty} \left| y_i \right| \left\| f_i \right\| \geq \left| \pi \right| \lim_{i \to \infty} \left| \pi y_i \right|$$

hence Ψ is well-defined and is clearly linear.

$$
\begin{aligned}
\left\| \Psi(y) \right\| &= \left\| (y_i)_{i \in \mathbb{N}} \right\| \\
&= \sup_{i \in \mathbb{N}} \left| y_i \right| \\
&\leq \frac{1}{\left| \pi \right|} \sup_{i \in \mathbb{N}} \left\| f_i \right\| \left| y_i \right|, \quad \text{as } \left| \pi \right| \leq \left\| f_i \right\| \\
&= \frac{1}{\left| \pi \right|} \left\| y \right\|
\end{aligned}
$$

hence Ψ is continuous. From their definitions, it is clear that Φ and Ψ are inverses of each other. This concludes the proof.

2.5 Bibliographical Notes

The material in this chapter mostly comes from the following sources: Diarra [20], Diagana [13], Schikhof [46], and van Rooij [53].

Chapter 3
Bounded Linear Operators in Non-Archimedean Banach Spaces

This chapter is devoted to basic properties of bounded linear operators on non-archimedean Banach spaces. The proofs of some of these basic results will be given. Special emphasis will be upon some of these classes of bounded linear operators including finite rank linear operators, completely continuous linear operators, and Fredholm linear operators.

In this chapter, $(\mathbb{K}, |\cdot|)$ and $(\mathbb{X}, \|\cdot\|)$ stand respectively for a non-trivial field which is complete with respect to a non-archimedean valuation $|\cdot|$ and a Banach space over the field $\mathbb{K}$. Further, the zero and identity operators will be denoted respectively by O and I, which are defined by,

$$I(x) = x \quad \text{and} \quad O(y) = 0$$

for all $x, y \in \mathbb{X}$.

3.1 Bounded Linear Operators

3.1.1 Definitions and Examples

Recall that a mapping $A : \mathbb{X} \mapsto \mathbb{X}$ is said to be a linear operator if, $A(\alpha x + \beta y) = \alpha Ax + \beta Ay$ for all $\alpha, \beta \in \mathbb{K}$ and $x, y \in \mathbb{X}$. A linear operator $A : \mathbb{X} \mapsto \mathbb{X}$ is said to be bounded if there exists $C \geq 0$ such that

$$\left\| Ax \right\| \leq C \left\| x \right\|$$

for all $x \in \mathbb{X}$.

© The Author(s) 2016
T. Diagana, F. Ramaroson, *Non-Archimedean Operator Theory*,
SpringerBriefs in Mathematics, DOI 10.1007/978-3-319-27323-5_3

Recall also that $B(\mathbb{X})$ denotes the collection of all bounded linear operators from $\mathbb{X}$ into itself. It is clear that if $A \in B(\mathbb{X})$, then the quantity, called the norm-operator of A,

$$\left\| A \right\| := \sup_{x \in \mathbb{X} \setminus \{0\}} \frac{\left\| Ax \right\|}{\left\| x \right\|}$$

is finite.

By definition of the norm-operator, if $A \in B(\mathbb{X})$, then the following identity holds,

$$\left\| Ax \right\| \leq \left\| A \right\| \cdot \left\| x \right\| \quad \text{for all } x \in \mathbb{X}. \tag{3.1}$$

Note that every bounded linear operator on $\mathbb{X}$ is continuous. Indeed, if $(x_n)_{n \in \mathbb{N}} \subset \mathbb{X}$ is a sequence which converges strongly to some $x \in \mathbb{X}$, that is, $\|x_n - x\| \to 0$ as $n \to \infty$, then using Eq. (3.1) it follows that $\|A(x_n - x)\| \leq \|A\| \cdot \|x_n - x\|$ which yields $\|A(x_n - x)\| \to 0$ as $n \to \infty$, that is, A is continuous. The converse, as given in the next theorem, is also true.

Theorem 3.1. *Every continuous linear operator* $A : \mathbb{X} \to \mathbb{X}$ *is bounded.*

Proof. Suppose A is continuous. Consequently, A is continuous at $x = 0$. Hence, there exists $\eta > 0$ such that $\|Ax\| \leq 1$ whenever $\|x\| \leq \eta$. Suppose the valuation of the non-archimedean field $\mathbb{K}$ is dense. Consequently, there exists $z_\eta \in \mathbb{K} \setminus \{0\}$ such that $|z_\eta| = \eta$. If $0 \neq x \in \mathbb{X}$, then let $z_x \in \mathbb{K} \setminus \{0\}$ such that $|z_x| = \|x\|$.

We have

$$\left\| \frac{z_\eta x}{z_x} \right\| = \eta.$$

Now

$$1 \geq \left\| A \left(\frac{z_\eta x}{z_x} \right) \right\| = \frac{|z_\eta| \, \|Ax\|}{|z_x|} = \frac{\eta \, \|Ax\|}{\|x\|},$$

and hence $\|Ax\| \leq \eta^{-1} \|x\|$ which yields A is bounded.

One should point out that the proof is similar in the case when the valuation of $\mathbb{K}$ is discrete and hence is omitted.

Example 3.2. Let $\mathbb{X} = \mathbb{K}^n = \{(x_1, x_2, \ldots, x_n) : z_k \in \mathbb{K}, \ k = 1, 2, \ldots, n\}$ be equipped with its natural non-archimedean norm given by

$$\left\| x \right\| = \max_{i=1,\ldots,n} |x_i|$$

for all $x = (x_1, x_2, \ldots, x_n) \in \mathbb{K}^n$.

Let $(e_1, e_2, \ldots, e_n)$ be the canonical basis of $\mathbb{K}^n$ defined by, $e_1 = (1, 0, 0, \ldots, 0)$, $e_2 = (0, 1, 0, \ldots, 0)$, ..., $e_n = (0, 0, \ldots, 1)$. Clearly, for all $x = (x_1, x_2, \ldots, x_n) \in \mathbb{K}^n$,

$$x = \sum_{j=1}^{n} x_j e_j \text{ for some } x_j \in \mathbb{F}, \ j = 1, 2, \ldots, n.$$

Let $A : \mathbb{K}^n \mapsto \mathbb{K}^n$ be a linear mapping. Clearly, $Ae_i \in \mathbb{K}^n$ and hence there exists $a_{ij} \in \mathbb{K}$ for $i, j = 1, 2, \ldots, n$ such that

$$Ae_j = \sum_{i=1}^{n} a_{ij} e_i.$$

In what follows, we show that the arbitrary linear operator A given above is necessarily bounded. Indeed, for all

$$x = \sum_{j=1}^{n} x_j e_j \text{ and } y = \sum_{j=1}^{n} y_j e_j,$$

$$\left\| Ax - Ay \right\| = \left\| \sum_{j=1}^{n} (x_j - y_j) Ae_j \right\|$$

$$\leq C \max \left(\left| x_1 - y_1 \right|, \left| x_2 - y_2 \right|, \ldots, \left| x_n - y_n \right| \right)$$

$$= C \left\| x - y \right\|,$$

where

$$C = \max_{j=1,\ldots,n} \left\| Ae_j \right\| = \max_{j=1,\ldots,n} \left(\max_{i=1,\ldots,n} \left| a_{ij} \right| \right) < \infty.$$

Consequently, $A : \mathbb{K}^n \mapsto \mathbb{K}^n$ is a bounded linear operator.

Example 3.3. Let $\mathbb{K} = (\mathbb{Q}_p, |\cdot|_p)$ where $p \geq 2$ is a prime and let $X = C(\mathbb{Z}_p, \mathbb{Q}_p)$ be the space of continuous functions from $\mathbb{Z}_p$ into $\mathbb{Q}_p$, which we equip with its sup-norm given by $\|f\|_\infty = \max_{z \in \mathbb{Z}_p} |f(z)|_p$ for all $f \in C(\mathbb{Z}_p, \mathbb{Q}_p)$. Let $P : \mathbb{Z}_p \mapsto \mathbb{Q}_p, z \mapsto P(z) = \sum_{k=1}^{N} a_k z^k$ be a polynomial of degree N with coefficients $a_1, a_2, \ldots, a_N$ belonging to $\mathbb{Q}_p$. Consider the so-called multiplication operator defined by, $A(f)(z) = P(z)f(z)$ for all $f \in C(\mathbb{Z}_p, \mathbb{Q}_p)$ and $z \in \mathbb{Z}_p$. Clearly, A is a linear operator. Further, $\|Af\|_\infty \leq C\|f\|_\infty$ for all $f \in C(\mathbb{Z}_p, \mathbb{Q}_p)$ where $C = \max(|a_1|_p, |a_2|_p, \ldots, |a_N|_p)$. Therefore, A is a bounded linear operator.

Example 3.4. Let $\mathbb{K} = (\mathbb{Q}_p, |\cdot|_p)$ where $p \geq 2$ is a prime. As in Example 3.3, let $\mathbb{X} = C(\mathbb{Z}_p, \mathbb{Q}_p)$ be the space of continuous functions from $\mathbb{Z}_p$ into $\mathbb{Q}_p$ be equipped with its usual sup-norm. Fix $\xi \in \mathbb{Z}_p$ and consider the operator defined by, $A_\xi(f)(z) = f(z + \xi) - f(z)$ for all $f \in C(\mathbb{Z}_p, \mathbb{Q}_p)$ and $z \in \mathbb{Z}_p$. It is clear that A_ξ is a linear operator. Further, $\|A_\xi f\|_\infty \leq \|f\|_\infty$ for all $f \in C(\mathbb{Z}_p, \mathbb{Q}_p)$. Therefore, A_ξ belongs to $B(C(\mathbb{Z}_p, \mathbb{Q}_p))$.

3.1.2   Basic Properties

Theorem 3.5. *If* $A, B \in B(\mathbb{X})$ *and* $\lambda \in \mathbb{K}$, *then,* $A + B$, λA, AB, *and* BA *all belong to* $B(\mathbb{X})$.

In view of the above, the space of bounded linear operators on $\mathbb{X}$ equipped with its above-mentioned operator-norm $\|\cdot\|$ is a normed vector space. In fact, as shown in the next theorem, $B(\mathbb{X})$ is a Banach space.

Theorem 3.6. *The space* $(B(\mathbb{X}), \|\cdot\|)$ *of bounded linear operator on* $\mathbb{X}$ *is a Banach space.*

Proof. The proof can be done slightly as in the classical setting. Indeed, let $(A_n)_{n \in \mathbb{N}}$ be a Cauchy sequence in $B(\mathbb{X})$. Equivalently, for all $\varepsilon > 0$ there exists $N \in \mathbb{N}$ such that $\|A_n - A_m\| < \varepsilon$ for all $n, m > N$. Our main task consists of proving that there exists a bounded linear operator $A : \mathbb{X} \mapsto \mathbb{X}$ such that $\|A_n - A\| \to 0$ as $n \to \infty$. For all $0 \neq x \in \mathbb{X}$, we have,

$$\left\| (A_n - A_m)x \right\| < \varepsilon \left\| x \right\| \tag{3.2}$$

for $n, m > N$.

Consequently, $(A_n x)_{n \in \mathbb{N}}$ is a Cauchy sequence in $\mathbb{X}$. Since $\mathbb{X}$ is a non-archimedean Banach space, there exists $\xi \in \mathbb{X}$ such that $\|A_n x - \xi\| \to 0$ as $n \to \infty$.

Setting

$$Ax := \xi = \lim_{n \to \infty} A_n x,$$

one defines a linear operator $A : \mathbb{X} \mapsto \mathbb{X}$.

Letting $m \to \infty$ in Eq. (3.2), one obtains, $\|A_n x - Ax\| \leq \varepsilon \|x\|$ for $n > N$. Consequently,

$$\left\| Ax \right\| = \left\| Ax - A_n x + A_n x \right\|$$

$$\leq \max \left(\left\| Ax - A_n x \right\|, \left\| A_n x \right\| \right)$$

$$\leq \max \left(\varepsilon \left\| x \right\|, \left\| A_n x \right\| \right)$$

$$\leq \max \left(\varepsilon \|x\|, \|A_n\| \|x\| \right)$$

$$= \max \left(\varepsilon, \|A_n\| \right) \|x\|$$

for $n > N$.

This yields A is a bounded operator. Further, $\|A_n - A\| \leq \varepsilon$ for $n > N$. Equivalently, $\|A_n - A\| \to 0$ as $n \to \infty$.

3.1.3 Bounded Linear Operators in Free Banach Spaces

Let $\mathbb{X}$ be a free Banach space over the non-archimedean field $(\mathbb{K}, |\cdot|)$ with canonical orthogonal basis $(e_j)_{j \in I}$. Define $e_i' \in \mathbb{X}^*$ by setting

$$x = \sum_{i \in I} x_i e_i, \quad e_i'(x) = x_i.$$

It turns out that $\|e_i'\|_* = \|e_i\|^{-1}$. Furthermore, every $x' \in \mathbb{X}^*$ can be expressed as $x' = \sum_{i \in I} \langle x', e_i \rangle e_i'$ with

$$\|x'\|_* = \sup_{i \in I} \frac{|\langle x', e_i \rangle|}{\|e_i\|}.$$

For each $f' \in \mathbb{X}^*$, define the linear operator $v' \otimes u : \mathbb{X} \mapsto \mathbb{X}$ by $(v' \otimes u)(w) := \langle v', w \rangle u$. Clearly, the operator $(v \otimes u)$ is bounded as $\|v' \otimes u\| = \|v'\|_* \|u\|$. Among other things, if $(e_i')_{i \in I}$ is the dual canonical orthogonal basis for $\mathbb{X}^*$, then $(e_i' \otimes e_j)_{(i,j) \in I \times I} \in B(\mathbb{X})$ and its operator-norm is given explicitly by

$$\|e_i' \otimes e_j\| = \frac{\|e_j\|}{\|e_i\|}.$$

Proposition 3.7 ([16]). *Let $A \in B(\mathbb{X})$, then it can be written in a unique fashion as a pointwise convergent series*

$$A = \sum_{(i,j) \in I \times I} a_{ji} e_i' \otimes e_j, \quad i \in I, \quad \lim_j |a_{ji}| \|e_j\| = 0. \tag{3.3}$$

Moreover,

$$\|A\| = \sup_{i \in I} \sup_{j \in I} \frac{|a_{ji}| \|e_j\|}{\|e_i\|}.$$

Proof. For all $j \in I$, $Ae_j = \sum_{i \in I} a_{ij}e_i$ where $a_{ij} \in \mathbb{K}$, $\lim_i |a_{ij}| \|e_j\| = 0$. Now for any $x = \sum_{j \in I} x_j e_j \in \mathbb{X}$,

$$Ax = \sum_{j \in I} \sum_{i \in I} a_{ij} x_j e_i = \sum_{j \in I} \sum_{i \in I} a_{ij} \left(e'_j \otimes e_i\right) x.$$

It remains to show that

$$\|A\| = \sup_{j \in I} \frac{\|Ae_j\|}{\|e_j\|} = \sup_{j \in I} \sup_{i \in I} \frac{|a_{ij}| \|e_e\|}{\|e_j\|}.$$

Indeed,

$$\frac{\|Ae_j\|}{\|e_j\|} \le \|A\|.$$

Next, for any $x = \sum_{j \in I} x_j e_j$,

$$\|Ax\| = \left\| \sum_{j \in I} x_j Ae_j \right\|$$

$$\le \sup_{j \in I} \left(|x_j| \cdot \|Ae_j\| \right)$$

$$= \sup_{j \in I} \left(|x_j| \cdot \|e_j\| \cdot \frac{\|Ae_j\|}{\|e_j\|} \right)$$

$$\le \|x\| \cdot \sup_{j \in I} \frac{\|Ae_j\|}{\|e_j\|}.$$

3.2   Additional Properties of Bounded Linear Operators

3.2.1   The Inverse Operator

Definition 3.8. A bounded linear operator $A : \mathbb{X} \mapsto \mathbb{X}$ is *invertible* if there exists $B \in B(\mathbb{X})$ such that $AB = BA = I$. The operator B is then called the inverse of A and denoted A^{-1}.

In the rest of the book, if $A \in B(\mathbb{X})$, we define its kernel $N(A)$ and its range $R(A)$ respectively by

$$N(A) := \left\{ x \in \mathbb{X} : Ax = 0 \right\}$$

and

$$R(A) := \left\{ Ax : x \in \mathbb{X} \right\}.$$

Theorem 3.9. *If $A \in B(\mathbb{X})$ such that $R(A) = \mathbb{X}$ and $N(A) = \{0\}$, then A^{-1} exists and belongs to $B(\mathbb{X})$.*

Proof. Obviously the operator A is invertible. The only thing we need to prove is the fact that A^{-1} is bounded. This in fact is guaranteed by the so-called open mapping theorem (see [53]). □

Theorem 3.10. *Let $A \in B(\mathbb{X})$ and suppose that $\|I - A\| < 1$. Then the following hold:*

(1) For any $x \in \mathbb{X}$, the series $\displaystyle\sum_{n=0}^{\infty} (I - A)^n x$ converges in $\mathbb{X}$ and $\displaystyle\sum_{n=0}^{\infty} (I - A)^n \in B(\mathbb{E}_\omega)$.

(2) A is invertible and $A^{-1} = \displaystyle\sum_{n=0}^{\infty} (I - A)^n$.

Proof. (1) Let $B = I - A$, so that $\|B\| < 1$. Let $x \in \mathbb{X}$, to show the convergence of the series $\displaystyle\sum_{n=0}^{\infty} B^n x$ we need to show that $\lim_{n \to \infty} B^n x = 0$. Note that $\|B^n x\| \leq \|B\|^n \|x\|$, but since $\|B\| < 1$, $\|B\|^n \|x\| \to 0$ as $n \to \infty$. Hence the series converges. It follows that the von Neumann series $\displaystyle\sum_{n=0}^{\infty} B^n$ is a well defined operator whose domain is $\mathbb{X}$. Moreover,

$$\left\| \sum_{n=0}^{\infty} B^n \right\| \leq \sup_{n \in \mathbb{N}} \left\| B^n \right\| \leq \sup_{n \in \mathbb{N}} \left\| B \right\|^n \leq 1,$$

and hence $\sum\limits_{n=0}^{\infty} B^n \in B(\mathbb{X})$.

(2) We simply observe that $A\left(\sum\limits_{n=0}^{\infty} B^n\right) = (I - B)\left(\sum\limits_{n=0}^{\infty} B^n\right) = I$, and, hence

$$A^{-1} = \sum_{n=0}^{\infty} B^n.$$

3.2.2   Perturbations of Orthogonal Bases Using the Inverse Operator

The main objective here is to make extensive use of the notion of inverse operator introduced in Sect. 3.2.1 to study perturbations of orthogonal bases of $\mathbb{E}_\omega$. Namely, if $(h_j)_{j \in \mathbb{N}}$ is an orthogonal basis and if $(f_j)_{j \in \mathbb{N}}$, is another sequence of vector of $\mathbb{E}_\omega$, not necessarily an orthogonal basis, such that the difference $f_j - h_j$ is small in a certain sense; we investigate conditions under which the family $(f_j)_{j \in \mathbb{N}}$ is a basis of $\mathbb{E}_\omega$.

We introduce the following notion of *strongly orthogonal basis*, which obviously is stronger than the notion or orthogonal basis introduced in Chap. 2 (see Definition 2.47).

Definition 3.11. A sequence $(h_j)_{j \in \mathbb{N}} \subset \mathbb{E}_\omega$ is called a *strongly orthogonal basis* if it satisfies the following:

(i) If $x = \sum\limits_{j \in \mathbb{N}} x_j h_j$, then $\|x\| = \sup\limits_{j \in \mathbb{N}} |x_j| \, \|h_j\|$;

(ii) $\langle h_i, h_j \rangle = \delta_{ij}$.

Theorem 3.12. *Let $(h_j)_{j \in \mathbb{N}}$ be a strongly orthogonal basis and let $(f_j)_{j \in \mathbb{N}}$ be a sequence of vectors in $\mathbb{E}_\omega$ satisfying the following condition: there exists $\alpha \in (0, 1)$ such that for every $x = \sum\limits_{j \in \mathbb{N}} x_j h_j \in \mathbb{E}_\omega$,*

$$\sup_{j \in \mathbb{N}} |x_j| \left\|f_j - h_j\right\| \leq \alpha \, . \, \sup_{j \in \mathbb{N}} |x_j| \left\|h_j\right\| = \alpha \, . \, \|x\|.$$

Then $(f_j)_{j \in \mathbb{N}}$ is a strongly orthogonal basis. Moreover, for any $y = \sum\limits_{j \in \mathbb{N}} y_j f_j \in \mathbb{E}_\omega$,

$$\|y\| = \sup_{j \in \mathbb{N}} \left(|y_j| \cdot \|f_j\| \right).$$

Proof. We first observe that, if $x = h_j$, then, the condition implies that for each j,

$$\left\| f_j - h_j \right\| \leq \alpha \left\| h_j \right\| < \left\| h_j \right\|$$

and hence $\left\| f_j \right\| = \left\| h_j \right\|$.

Next, for any $x = \sum_{j \in \mathbb{N}} x_j h_j \in \mathbb{E}_\omega$, $\left| x_j \right| \left\| f_j \right\| = \left| x_j \right| \left\| h_j \right\|$, that is, $\lim_{j \to \infty} \left(\left| x_j \right| \cdot \left\| f_j \right\| \right)$
$= 0$. Therefore, the operator defined by

$$Ax = \sum_{j \in \mathbb{N}} x_j f_j$$

is well-defined and satisfies $Ah_j = f_j$. Moreover

$$\left\| x - Ax \right\| = \left\| \sum_{j \in \mathbb{N}} x_j \left(h_j - f_j \right) \right\|$$

$$\leq \sup_{j \in \mathbb{N}} \left(\left| x_j \right| \cdot \left\| h_j - f_j \right\| \right)$$

$$\leq \alpha \cdot \sup_{j \in \mathbb{N}} \left(\left| x_j \right| \cdot \left\| h_j \right\| \right)$$

$$= \alpha \cdot \left\| x \right\|.$$

It follows that $\| I - A \| \leq \alpha < 1$, and hence A is invertible, by Theorem 3.10. It remains to show that A is isometric.

In the above, observe that the inequalities: $\| x - Ax \| \leq \alpha \| x \| < \| x \|$ imply that $\| Ax \| = \| x \|$ for any $x \in \mathbb{E}_\omega$, hence, A is isometric. Consequently, $(f_j)_{j \in \mathbb{N}}$ is a strongly orthogonal basis. Moreover, for any $y = \sum_{j \in \mathbb{N}} y_j f_j \in \mathbb{E}_\omega$

$$\left\| y \right\| = \left\| A(\sum_{j \in \mathbb{N}} y_j h_j) \right\| = \left\| \sum_{j \in \mathbb{N}} y_j h_j \right\| = \sup_{j \in \mathbb{N}} \left(\left| y_j \right| \cdot \left\| h_j \right\| \right).$$

Theorem 3.13. *Let* $(h_j)_{j \in \mathbb{N}}$ *be a strongly orthogonal basis,* $C \in B(\mathbb{E}_\omega)$ *invertible such that*

$$\left\| C^{-1} \right\| = \left\| C \right\|^{-1}.$$

Suppose that $(f_j)_{j \in \mathbb{N}}$ is a sequence of vectors in $\mathbb{E}_\omega$ satisfying the following condition

$$\sup_{j \in \mathbb{N}} \frac{\left\| f_j - C h_j \right\|}{\left\| h_j \right\|} < \left\| C \right\|,$$

then $(f_j)_{j \in \mathbb{N}}$ is a strongly orthogonal basis.

Proof. We first observe that for any j, $\left\| f_j \right\| \leq \left\| C \right\| \left\| h_j \right\|$. For any $x = \sum_{j \in \mathbb{N}} x_j h_j \in \mathbb{E}_\omega$:

$$\lim_{j \to \infty} \left| x_j \right| \left\| f_j \right\| = \lim_{j \to \infty} \left| x_j \right| \left\| h_j \right\| \frac{\left\| f_j \right\|}{\left\| h_j \right\|}$$

$$\leq \left\| C \right\| \cdot \lim_{j \to \infty} \left| x_j \right| \left\| h_j \right\|$$

$$= 0.$$

Therefore if we put $Ax = \sum_{j \in \mathbb{N}} x_j f_j$, then, A is a well-defined operator satisfying $A h_j = f_j$. The second condition of the theorem implies that if we put $B = C - A$, then,

$$\left\| B \right\| = \left\| C - A \right\| = \left\| A - C \right\| < \left\| C \right\|$$

from which we deduce that $\|A\| = \|C\|$. Next,

$$\left\| BC^{-1} \right\| \leq \left\| B \right\| \left\| C^{-1} \right\| < \left\| C \right\| \left\| C^{-1} \right\| = 1,$$

by assumption.

Therefore the operator AC^{-1} is such that $\left\| 1 - AC^{-1} \right\| = \left\| BC^{-1} \right\| < 1$. We can apply Theorem 3.10 to AC^{-1} and find that it is invertible. Since C is also invertible, it follows that A is invertible, and, hence, $(f_j)_{j \in \mathbb{N}}$ is a strongly orthogonal basis for $\mathbb{E}_\omega$.

Corollary 3.14. *Let $(h_j)_{j \in \mathbb{N}} \subset \mathbb{E}_\omega$ be a strongly orthogonal basis, $(f_j)_{j \in \mathbb{N}} \subset \mathbb{E}_\omega$ a sequence of vectors and ζ a non-zero element of $\mathbb{K}$ satisfying:*

$$\sup_{j \in \mathbb{N}} \frac{\left\| f_j - \zeta h_j \right\|}{\left\| h_j \right\|} < \left| \zeta \right|,$$

then, $(f_j)_{j \in \mathbb{N}}$ is a strongly orthogonal basis for $\mathbb{E}_\omega$.

Proof. In Theorem 3.13, we take the matrix C to be the diagonal matrix

$$C = \sum_{i,j \in \mathbb{N}} c_{ij} \left(h'_j \otimes h_i \right)$$

with $c_{ij} = 0$ if $i \neq j$ and $c_{ii} = \zeta$ for all $i \geq 0$.

It is clear that for any j, $Ch_j = \zeta h_j$, C is invertible, $C^{-1}h_j = \zeta^{-1}h_j$, $\|C\| = |\zeta|$ and $\|C^{-1}\| = |\zeta|^{-1}$. Moreover

$$\sup_{j \in \mathbb{N}} \frac{\left\| f_j - Ch_j \right\|}{\left\| h_j \right\|} = \sup_{j \in \mathbb{N}} \frac{\left\| f_j - \zeta h_j \right\|}{\left\| h_j \right\|}$$

$$< |\zeta|$$

$$= \|C\|.$$

Corollary 3.15. *Let $(h_j)_{j \in \mathbb{N}}$ be a strongly orthogonal basis for $\mathbb{E}_\omega$. If $(g_j)_{j \in \mathbb{N}}$ is a sequence of vectors of $\mathbb{E}_\omega$ satisfying:* $\lim_{j \to \infty} \frac{|\langle h_i, g_j \rangle|}{|\omega_j|} \|g_j\| = 0$ *for each $i \in \mathbb{N}$, and that,*

$$\sup_{j \in \mathbb{N}} \left(\frac{\left\| h_j - Sh_j \right\|}{|\omega_j|^{1/2}} \right) < 1,$$

where $Sh_i = \sum_{j \in \mathbb{N}} \frac{\langle h_i, g_j \rangle}{\omega_j} g_j$ for each $i \in \mathbb{N}$, then $(Sh_j)_{j \in \mathbb{N}}$ is a strongly orthogonal basis of $\mathbb{E}_\omega$.

Remark 3.16.   (i) Note that S defined above is a linear operator on $\mathbb{E}_\omega$.

(ii) The assumption, $\lim_{j \to \infty} \frac{|\langle h_i, g_j \rangle|}{|\omega_j|} \|g_j\| = 0$ implies that $Sh_i = \sum_{j \in \mathbb{N}} \frac{\langle h_i, g_j \rangle}{\omega_j} g_j$, $i \in$ $\mathbb{N}$, is well-defined.

(iii) The operator S is isometric, by $\sup_{j \in \mathbb{N}} \left(\frac{\left\| h_j - Sh_j \right\|}{|\omega_j|^{1/2}} \right) < 1$.

Proof. It suffices to put $C = I$, the identity operator of $\mathbb{E}_\omega$, and $f_j = Sh_j$ in Theorem 3.13.

Example 3.17. We illustrate Corollary 3.14 with the following example: Let p be a prime, $\mathbb{K} = \mathbb{Q}$, $\omega_j = p^j$, $|\omega_j| = \dfrac{1}{p^j}$. As an orthogonal basis for $\mathbb{E}_\omega$ we use the canonical orthogonal basis $(e_j)_{j \in \mathbb{N}}$ and recall that $\|e_j\| = |\omega_j|^{1/2} = \dfrac{1}{p^{j/2}}$.

Let ς be such that $|\varsigma| = 1$ and let

$$g_j = (u - \varsigma)\, e_j + p^{1+j} \sum_{i \in \mathbb{N}, i \neq j} e_i$$

for each $s \in \mathbb{N}$.

We choose u such that $|u| = 1$ and $|u - \varsigma| = \dfrac{1}{p^{1+j}}$. To achieve this choice of u we do the following: ς is a p-adic unit which can be written in $\mathbb{K}$ as

$$\varsigma = a_0 + a_1 p + a_2 p^2 + \ldots + a_j p^j + a_{j+1} p^{j+1} + \ldots,$$

where $1 \le a_0 \le p - 1$ and for $k \neq 0$, $0 \le a_k \le p - 1$.

Put $u = a_0 + a_1 p + a_2 p^2 + \ldots + a_j p^j$, then $|u| = 1$ and

$$u - \varsigma = -(a_{j+1} p^{j+1} + \ldots),$$

hence $|u - \varsigma| = \dfrac{1}{p^{1+j}}$.

Next,

$$\|g_j\| = \max\left(|u - \varsigma| |\omega_j|^{1/2}, \frac{1}{p^{1+j}} \sup_{i \neq j} |\omega_i|^{1/2} \right)$$

$$= \max\left(\frac{1}{p^{1+j+j/2}}, \sup_{i \neq j} \frac{1}{p^{1+j+i/2}} \right)$$

$$= \max\left(\frac{1}{p^{1+j+j/2}}, \frac{1}{p^{1+j}} \right)$$

$$= \frac{1}{p^{1+j}} \text{ (even if } j = 0).$$

Hence,

$$\frac{\|g_j\|}{|\omega_j|^{1/2}} = \frac{1}{p^{1+j/2}},$$

and $\sup_{j} \dfrac{\|g_j\|}{|\omega_j|^{1/2}} = \sup_{j} \dfrac{1}{p^{1+j/2}} = \dfrac{1}{p} < |\varsigma| = 1.$ Finally, let $f_j = g_j + \zeta e_j = u e_j +$

$p^{1+j} \sum\limits_{i \neq j} e_i,$ then

$$\sup_{j \in \mathbb{N}} \frac{\|f_j - \zeta e_j\|}{\|e_j\|} = \sup_{j \in \mathbb{N}} \frac{\|g_j\|}{|\omega_j|^{1/2}} < |\zeta|.$$

3.2.3   The Adjoint Operator

As in the classical case, if $A \in B(\mathbb{E}_\omega)$, an adjoint of A is an operator A^* satisfying $\langle Ax, y \rangle = \langle x, A^*y \rangle$ for any x, y in $\mathbb{E}_\omega$. If it exists, the adjoint A^* is unique and has the same norm as A, and hence, lies in $B(\mathbb{E}_\omega)$ as well. The properties of the adjoint are easier to express in terms of $(e_s)_{s \in \mathbb{N}}$, the canonical orthogonal basis of $\mathbb{E}_\omega$.

Proposition 3.18 ([13, 16]). *Let $A = \sum\limits_{t \in \mathbb{N}} \sum\limits_{s \in \mathbb{N}} a_{st} \left(e'_t \otimes e_s \right) \in B(\mathbb{E}_\omega)$, then the adjoint A^* exists if and only if for all s,* $\lim\limits_{t \to \infty} \dfrac{|a_{st}|}{|\omega_t|^{1/2}} = 0.$ *In that event,*

$$A^* = \sum_{t \in \mathbb{N}} \sum_{s \in \mathbb{N}} \omega_s^{-1} \omega_t a_{ts} \left(e'_t \otimes e_s \right).$$

Proof. Let $A^* = \sum\limits_{t \in \mathbb{N}} \sum\limits_{s \in \mathbb{N}} b_{st} \left(e'_t \otimes e_s \right).$ It is clear that a linear operator A^* is the adjoint of A if and only if $\langle Ae_s, e_t \rangle = \langle e_s, A^*e_t \rangle$, that is,

$$\left\langle \sum_{k \in \mathbb{N}} a_{ks} e_k, e_t \right\rangle = a_{ts} \omega_t = \left\langle e_s, \sum_{k \in \mathbb{N}} b_{kt} e_k \right\rangle = b_{st} \omega_s, \quad \forall s, t \in \mathbb{N},$$

which yields $b_{st} = \omega_s^{-1} \omega_t a_{ts}$ for all $s, t \in \mathbb{N}$. Further, for all t,

$$\lim_{s \to \infty} \left| b_{st} \right| \left| \omega_s \right|^{1/2} = 0,$$

which is equivalent to $\lim\limits_{t \to \infty} \dfrac{|a_{st}|}{|\omega_t|^{1/2}} = 0,$ for all $s \in \mathbb{N}$. The proof is complete. $\quad\square$

Remark 3.19. It is possible that an operator in $B(\mathbb{E}_\omega)$ may fail to have an adjoint as shown in the following example: Let p be a prime and $\mathbb{K} = \mathbb{Q}_p$, $\omega_s = 1 + p^s$,

$$a_{st} = \begin{cases} p^s & \text{if } s < t \\ 0 & \text{if } s \geq t \end{cases}$$

and let $A = \sum_{s,t \in \mathbb{N}} a_{st} \left(e'_t \otimes e_s \right)$. Then

$$|a_{st}| = \begin{cases} p^{-s} & \text{if } s < t \\ 0 & \text{if } s \geq t \end{cases}$$

and $|\omega_s| = 1$. For all $t \in \mathbb{N}$, $\lim_{s \to \infty} |a_{st}| \, |\omega_s|^{1/2} = 0$. Moreover,

$$\|A\| = \sup_{s,t \in \mathbb{N}} \frac{\left| a_{st} \right| \left| \omega_s \right|^{1/2}}{\left| \omega_s \right|^{1/2}} = \sup_{s < t} p^{-s} = 1,$$

so that $A \in B(\mathbb{E}_\omega)$. However, for all s, $\lim_{t \to \infty} \frac{|a_{st}|}{|\omega_t|^{1/2}} = \lim_{t \to \infty} p^{-s} = p^{-s} \neq 0$, and

therefore, $A \notin B_0(\mathbb{E}_\omega)$.

In the rest of this book, $B_0(\mathbb{E}_\omega)$ stands for the collection of all bounded linear operators on $\mathbb{E}_\omega$ whose adjoint operators do exist. Namely,

$$B_0(\mathbb{E}_\omega) := \left\{ \sum_{i,j \in \mathbb{N}} a_{ij} \, e'_j \otimes e_i \in B(\mathbb{E}_\omega) : \lim_{j \to \infty} \frac{|a_{ij}|}{|\omega_j|^{1/2}} = 0, \ \forall i \in \mathbb{N} \right\}.$$

Proposition 3.20. *If $A, B \in B_0(\mathbb{E}_\omega)$ and for $\lambda \in \mathbb{K}$, then the following hold:*

(a) $(A + B)^* = A^* + B^*$;
(b) $(AB)^* = B^* A^*$;
(c) $(\lambda A)^* = \lambda A^*$;
(d) $(A^*)^* = A$;
(e) $\|A\| = \|A^*\|$.

Definition 3.21. An operator $A \in B_0(\mathbb{E}_\omega)$ is said to be self-adjoint if $A = A^*$.

Definition 3.22. An operator $A \in B_0(\mathbb{E}_\omega)$ is said to be normal if $AA^* = A^*A$.

Definition 3.23. An operator $A \in B_0(\mathbb{E}_\omega)$ is said to be unitary if $AA^* = A^*A = I$.

Remark 3.24. It should be mentioned that normal and unitary linear operators in the non-archimedean context were recently studied by Kochubei, see [34, 36].

3.3 Finite Rank Linear Operators

In this section, we introduce and study properties of the so-called finite rank operators in the non-archimedean context.

3.3.1 Basic Definitions

Definition 3.25. An operator $A \in B(\mathbb{X})$ is called an operator of finite rank if $R(A)$ is a finite dimensional subspace of $\mathbb{X}$.

Definition 3.26. If $A \in B(\mathbb{X})$ is a finite rank operator, then the dimension of $R(A)$, that is, $\dim R(A)$ is called the rank of A.

Definition 3.27. The collection of all finite rank operators from $\mathbb{X}$ into $\mathbb{X}$ will be denoted by $\mathcal{F}(\mathbb{X})$.

3.3.2 Properties of Finite Rank Operators

Proposition 3.28. *An operator $A \in \mathcal{F}(\mathbb{X})$ if and only if there exist finite sequences $u_1, u_2, \ldots, u_r \in \mathbb{X}$ and $u'_1, u'_2, \ldots, u'_r \in \mathbb{X}^*$ such that*

$$A = \sum_{k=1}^{r} u'_k \otimes u_k. \tag{3.4}$$

Proof. If there exist finite sequences $u_1, u_2, \ldots, u_r \in \mathbb{X}$ and $u'_1, u'_2, \ldots, u'_r \in \mathbb{X}^*$ such that Eq. (3.4) holds, then $A \in \mathcal{F}(\mathbb{X})$. Indeed, A is linear and $R(A) \subset \mathbb{X}$ is spanned by $\{u_1, u_2, \ldots, u_r\}$ and hence is of dimension at most r. Furthermore,

$$\left\| Ax \right\| \leq \max \left(\left\| u'_1 \right\|_* \cdot \left\| u_1 \right\|, \left\| u'_2 \right\|_* \cdot \left\| u_2 \right\|, \ldots, \left\| u'_r \right\|_* \cdot \left\| u_r \right\| \right) \left\| x \right\|$$

and hence A is bounded. Consequently, $A \in \mathcal{F}(\mathbb{X})$.

Conversely, let $A \in \mathcal{F}(\mathbb{X})$ and let $u_1, u_2, \ldots, u_r \in \mathbb{X}$ be a linearly independent system that spans $R(A)$. Then, for all $x \in \mathbb{X}$,

$$Ax = \sum_{k=1}^{r} \xi_k(x) u_k,$$

where $\xi_1, \ldots, \xi_r : \mathbb{X} \mapsto \mathbb{K}$ are linear functionals.

Clearly, each of the linear functionals ξ_k is bounded. To prove that, we will make use of the fact that all norms in a non-archimedean finite dimensional vector space are equivalent (Proposition 2.27). Consequently, there exists $C > 0$ such that

$$\max \left(\left\| \xi_1 \right\|_*, \left\| \xi_2 \right\|_*, \ldots, \left\| \xi_r \right\|_* \right) \leq C \max \left(\left| \xi_1(u_1) \right|, \left| \xi_2(u_2) \right|, \ldots, \left| \xi_r(u_r) \right| \right)$$

which yields

$$\max \left(\left| \xi_1(x) \right|, \left| \xi_2(x) \right|, \ldots, \left| \xi_r(x) \right| \right) \leq C \left\| Ax \right\| \leq C \left\| A \right\| \cdot \left\| x \right\|.$$

Consequently, there exist functionals $u'_k \in \mathbb{X}^*$ such that $\xi_k(x) = u'_k(x)$ which in turn, yields

$$Ax = \sum_{k=1}^{r} u'_k(x) u_k = \sum_{k=1}^{r} (u'_k \otimes u_k)(x)$$

and therefore, Eq. (3.4) holds.

Proposition 3.29. *If $A \in B(\mathbb{X})$ and $B \in \mathcal{F}(\mathbb{X})$, then AB and BA belong to $\mathcal{F}(\mathbb{X})$.*

Proof. To see that, write $B = \sum_{k=1}^{r} u'_k \otimes u_k$ where $u_k \in \mathbb{X}$, $u'_k \in \mathbb{X}^*$ for $k = 1, 2, \ldots, r$.

Now

$$AB = A \sum_{k=1}^{r} u'_k \otimes u_k = \sum_{k=1}^{r} A(u'_k \otimes u_k) = \sum_{k=1}^{r} (u'_k \otimes Au_k) \in \mathcal{F}(\mathbb{X}).$$

Similarly, letting $B = \sum_{k=1}^{r} u'_k \otimes u_k$ where $u_k \in \mathbb{X}$, $u'_k \in \mathbb{X}^*$ for $k = 1, 2, \ldots, r$ it follows that,

$$BA = \sum_{k=1}^{r} (u'_k \otimes u_k)A = \sum_{k=1}^{r} (u'_k A) \otimes u_k \in \mathcal{F}(\mathbb{X}).$$

3.4 Completely Continuous Linear Operators

3.4.1 Basic Properties

Definition 3.30. A bounded linear operator $K : \mathbb{X} \mapsto \mathbb{X}$ is said to be completely continuous if there exists a sequence $F_n \in \mathcal{F}(\mathbb{X})$ such that $\|K - F_n\| \to 0$ as $n \to \infty$. The collection of such linear operators will be denoted $C(\mathbb{X})$.

Example 3.31. Classical examples of completely continuous operators include finite rank operators.

Theorem 3.32. *If $A, B \in B(\mathbb{X})$ are completely continuous linear operators, then*

(1) $A + B$ is completely continuous;
(2) If $C \in B(\mathbb{X})$ and $D \in B(\mathbb{X})$, then AC and DA are completely continuous.

Proof. (1) Let $A_n, B_n \in \mathcal{F}(\mathbb{X})$ such that $\|A - A_n\| \to 0$ and $\|B - B_n\| \to 0$ as $n \to \infty$. Using the facts that $A_n + B_n \in \mathcal{F}(\mathbb{X})$ and that

$$\left\| (A + B) - (A_n + B_n) \right\| \to 0 \ \text{ as } \ n \to \infty,$$

we deduce that $A + B \in C(\mathbb{X})$.
(2) Let $A_n \in \mathcal{F}(\mathbb{X})$ such that $\|A - A_n\| \to 0$ as $n \to \infty$. Since $\dim R(A_n) < \infty$ and $R(A_n C) \subset R(A_n)$ for all $n \in \mathbb{N}$ it follows that $A_n C \in \mathcal{F}(\mathbb{X})$.
Now

$$\left\| AC - A_n C \right\| \le \left\| A_n - A \right\| \cdot \left\| C \right\| \to 0$$

as $n \to \infty$ and hence $AC \in C(\mathbb{X})$.
Similarly, according to Proposition 3.29, $DA_n \in \mathcal{F}(\mathbb{X})$ for all $n \in \mathbb{N}$.
Now

$$\left\| DA - DA_n \right\| \le \left\| A_n - A \right\| \cdot \left\| D \right\| \to 0$$

as $n \to 0$ and hence $DA \in C(\mathbb{X})$.

3.4.2 Completely Continuous Linear Operators on $\mathbb{E}_\omega$

The following characterization of completely continuous operators on $\mathbb{E}_\omega$ is due to Diarra [21].

Proposition 3.33. *Let $A = \sum\limits_{i,j \in \mathbb{N}} a_{ij} \left(e'_j \otimes e_i \right) \in B(\mathbb{E}_\omega)$. The operator A is completely continuous if and only if*

$$\limsup_{\substack{i \to \infty \\ j \in \mathbb{N}}} \frac{|a_{ij}|}{\|e_j\|} \|e_i\| = 0.$$

Example 3.34. Suppose that $\mathbb{X} = \mathbb{E}_\omega$ is equipped with its natural topology. Consider the diagonal operator D defined by, $De_j = \lambda_j e_j$ where the sequence $\lambda_j \in \mathbb{K}$ for each $j \in \mathbb{N}$, satisfies:

$$\lim_{j \to \infty} |\lambda_j| = 0.$$

Under the previous condition, we claim that D is completely continuous. Indeed, consider the sequence of linear operators D_n defined on $\mathbb{E}_\omega$ by

$$D_n e_j = \lambda_j e_j \text{ for } j = 0, 1, \ldots, n \text{ and } D_n e_j = 0 \text{ for } j \geq n + 1.$$

Clearly, $D_n \in \mathcal{F}(\mathbb{E}_\omega)$ as its image $R(D_n)$ is a subspace of the space generated by $\{e_j : j = 0, 1, \ldots, n\}$. Moreover,

$$\lim_{n \to \infty} \|D - D_n\| = \lim_{n \to \infty} \sup_{j \geq n+1} |\lambda_j| = 0,$$

and hence $D \in C(\mathbb{E}_\omega)$.

Remark 3.35. As we have previously seen, if $A, B \in C(\mathbb{E}_\omega)$ and $\lambda \in \mathbb{K}$, then $A + \lambda B$ and AB belongs to $C(\mathbb{E}_\omega)$.

Let $C_0(\mathbb{E}_\omega)$ be the collection of all completely continuous linear operators on $\mathbb{E}_\omega$ that admit an adjoint. If $A \in C_0(\mathbb{E}_\omega)$ and $B \in B_0(\mathbb{E}_\omega)$, then the linear operators A^*, AB, and BA belong to $\in C_0(\mathbb{E}_\omega)$.

For more on completely continuous linear operators in the non-archimedean setting and related issues, we refer the reader to Serre [48].

3.5 Bounded Fredholm Linear Operators

3.5.1 Definitions and Examples

Definition 3.36. An operator $A \in B(\mathbb{X})$ is said to be a Fredholm operator if it satisfies the following conditions:

(a) $\eta(A) := \dim N(A)$ is finite;
(b) $R(A)$ is closed; and
(c) $\delta(A) := \dim(\mathbb{X}/R(A))$ is finite.

The collection of all the above-mentioned Fredholm linear operators will be denoted by $\Phi(\mathbb{X})$. If $A \in \Phi(\mathbb{X})$, we then define its index by setting,

$$\chi(A) := \eta(A) - \delta(A).$$

Definition 3.37. The collection of all Fredholm linear operators $A \in B(\mathbb{X})$ will be denoted by $\Phi(\mathbb{X})$. If $A \in \Phi(\mathbb{X})$, we then define its index by setting,

$$\chi(A) := \eta(A) - \delta(A).$$

Example 3.38. Let $\mathbb{X}$ be a finite dimensional vector space. Then any linear operator $A : \mathbb{X} \mapsto \mathbb{X}$ is a Fredholm operator with index $\chi(A) := \eta(A) - \delta(A) = 0$.

Example 3.39. Suppose $\mathbb{X} = \mathbb{E}_\omega$. Any invertible bounded linear operator $A : \mathbb{E}_\omega \mapsto \mathbb{E}_\omega$ (in particular, the identity operator $I : \mathbb{E}_\omega \mapsto \mathbb{E}_\omega$, $I(u) = u$), is a Fredholm operator with index $\chi(A) = 0$ as $\eta(A) = \delta(A) = 0$.

Example 3.40. Let $S : l^2(\mathbb{N}) \mapsto l^2(\mathbb{N})$ be the unilateral shift defined by $Se_j = e_{j+1}$ for all $n \in \mathbb{N}$ where $(e_j)_{j \in \mathbb{N}}$ is the canonical basis of $l^2(\mathbb{N})$ defined by $e_1 = (1, 0, 0, \ldots)$, $e_2 = (0, 1, 0, 0, \ldots)$ and $e_n = (0, 0, \ldots 0, 0, 1, 0, 0, \ldots)$. It can be shown that S is a Fredholm bounded linear operator whose index is -1.

3.5.2 Properties of Fredholm Operators

In this subsection we follow the work of Gohberg et al. [27] on Fredholm bounded linear operators.

Let $A \in B(\mathbb{X})$. We suppose the bounded linear operator A has the property:

$N(A), R(A)$ are complemented by the (closed) subspaces $\mathbb{X}_0, \mathbb{Y}_0$.

Clearly, the product space $\mathbb{X}_0 \times \mathbb{Y}_0$ is a non-archimedean Banach space when it is equipped with the norm defined by

$$\left\| (x_0, y_0) \right\| := \max \left(\left\| x_0 \right\|, \left\| y_0 \right\| \right)$$

for all $(x_0, y_0) \in \mathbb{X}_0 \times \mathbb{Y}_0$.

Define the linear operator $\tilde{A} : \mathbb{X}_0 \times \mathbb{Y}_0 \mapsto \mathbb{X}$ by setting

$$\tilde{A}(x_0, y_0) = Ax_0 + y_0.$$

It is clear that the operator $\tilde{A}$ defined above is a bijective bounded linear operator. In what follows we say that $\tilde{A}$ is the bijection associated with the bounded linear operator A and the subspaces $\mathbb{X}_0$ and $\mathbb{Y}_0$.

If $A \in \Phi(\mathbb{X})$, then it can be easily seen that $\tilde{A} \in \Phi(\mathbb{X})$ with Y_0 being a finite dimensional subspace. Identifying $\mathbb{X}_0$ with $\mathbb{X}_0 \times \{0\}$ it follows that the linear operator defined by $A_0 : \mathbb{X}_0 \mapsto \mathbb{X}$, $Ax_0 = Ax$, is the restriction of both A and $\tilde{A}$.

Lemma 3.41. *Suppose $A_0 : L \mapsto \mathbb{X}$ is a restriction of $A \in B(\mathbb{X})$ to a subspace $L \subset \mathbb{X}$ with codim $L = N < \infty$. Then A is Fredholm if and only if A_0 is Fredholm. In this event, $\chi(A) = \chi(A_0) + N$.*

The proof of this lemma is similar to the classical case (see [27, Lemma 3.1]) and hence is omitted.

Theorem 3.42. *If $A, B : \mathbb{X} \mapsto \mathbb{X}$ are Fredholm linear operators, then so is their composition $BA : \mathbb{X} \mapsto \mathbb{X}$, and*

$$\chi(BA) = \chi(B) + \chi(A).$$

Proof. Let $\tilde{A}$ be the bijection associated with A and the subspaces $\mathbb{X}_0$ and $\mathbb{Y}_0$. Further, let A_0 be the restriction of both A to $\mathbb{X}_0$. Consider the linear operator $B\tilde{A}$. Since the linear operator $\tilde{A}$ is bijective it follows that $B\tilde{A}$ is Fredholm with $\chi(B\tilde{A}) = \chi(B)$. It is also clear that BA_0 is a restriction of both $B\tilde{A}$ and BA. Using Lemma 3.41 it follows that

$$\chi(BA) = \chi(BA_0) + \dim(\mathbb{X}/\mathbb{X}_0)$$
$$= \chi(B\tilde{A}) + \dim\left(\mathbb{X}_0 \times \mathbb{Y}_0/\mathbb{X}_0 \times \{0\}\right) + \eta(A)$$
$$= \chi(A) + \chi(B).$$

Theorem 3.43. *Let $A : \mathbb{X} \mapsto \mathbb{X}$ be a Fredholm operator. If $B : \mathbb{X} \mapsto \mathbb{X}$ is a bounded linear operator such that $\|B\| < \|\tilde{A}^{-1}\|^{-1}$ where $\tilde{A}$ is the bijection associated with A, then $A + B$ is a Fredholm operator with $\chi(A + B) = \chi(A)$.*

Proof. Let $S = A + B$ and define the operator $\tilde{S} : \mathbb{X}_0 \times \mathbb{Y}_0 \mapsto \mathbb{X}$ defined by $\tilde{S}(x_0, y_0) = Sx_0 + y_0$. Using the fact that $\tilde{A}$ is bijective and that $\|\tilde{A} - \tilde{S}\| \leq \|A - S\| = \|B\| < \|\tilde{A}^{-1}\|^{-1}$ it follows that $\tilde{S}$ is bijective. Using Lemma 3.41 it follows that S is a Fredholm operator and that

$$\chi(S) = \chi(S_0) + \eta(A)$$
$$= \chi(\tilde{S}) - \delta(A) + \eta(A)$$
$$= \chi(A).$$

Lemma 3.44. *If $L \in B_0(\mathbb{E}_\omega)$, then $\overline{R(L)} \subset N(L^*)^\perp$.*

Proof. If $x \in N(L^*)$, then $L^*x = 0$ which yields $\langle y, L^*x \rangle = 0$ for all $y \in \mathbb{E}_\omega$, that is, $\langle Ly, x \rangle = 0$ for all $y \in \mathbb{E}_\omega$, which yields $x \in R(L)^\perp$. Now $N(L^*) \subset R(L)^\perp$ yields $\overline{R(L)} \subset N(L^*)^\perp$.

Question 3.45. Find conditions under which the identity $\overline{R(L)} = N(L^*)^\perp$ holds?

Proposition 3.46. *If $K \in C_0(\mathbb{E}_\omega)$, then $I - K$ belongs to $\Phi(\mathbb{E}_\omega)$.*

Proof. Clearly, $R(I - K)$ is closed. Using the fact that $x = Kx$ for each $x \in N(I - K)$ it follows that the identity operator is a completely continuous operator on $N(I - K)$ and hence one must have $\dim N(I - K) = \eta(I - K) < \infty$. Now since $K \in C_0(\mathbb{E}_\omega)$ it follows that $K^* \in C_0(\mathbb{E}_\omega)$. Now $R(I - K) \subset N(I - K^*)^\perp$ (see Lemma 3.44) and hence $\delta(I - K) \leq \dim N(I - K^*) < \infty$.

3.6 Spectral Theory for Bounded Linear Operators

This section is devoted to the spectral theory of bounded linear operators on a non-archimedean Banach space $(\mathbb{E}, \| \cdot \|)$.

3.6.1 The Spectrum of a Bounded Linear Operator

Definition 3.47. Let $(\mathbb{E}, \| \cdot \|)$ be a non-archimedean Banach space. The resolvent set of a bounded linear operator $A : \mathbb{E} \mapsto \mathbb{E}$ is defined by

$$\rho(A) := \left\{ \lambda \in \mathbb{K} : \lambda I - A \text{ is invertible in } B(\mathbb{E}) \right\}.$$

Using the open mapping theorem (see [53]), it follows that the resolvent set $\rho(A)$ of A can be reformulated as follows,

$$\rho(A) := \left\{ \lambda \in \mathbb{K} : R_\lambda^A := (\lambda I - A)^{-1} \in B(\mathbb{E}) \right\}.$$

The function $\lambda \mapsto R_\lambda^A := (\lambda I - A)^{-1}$ is called the resolvent of A.

Definition 3.48. The spectrum $\sigma(A)$ of a bounded linear operator $A : \mathbb{E} \mapsto \mathbb{E}$ is defined by $\sigma(A) = \mathbb{K} \setminus \rho(A)$.

Definition 3.49. A scalar $\lambda \in \mathbb{K}$ is called an eigenvalue of $A \in B(\mathbb{E})$ whenever there exists a nonzero $u \in \mathbb{E}$ (called eigenvector associated with λ) such that $Au = \lambda u$.

In view of Definition 3.49, eigenvalues of $A \in B(\mathbb{E})$ consist of all $\lambda \in \mathbb{K}$ for which $\lambda I - A$ is not one-to-one, that is, $N(\lambda I - A) \neq \{0\}$. The collection of all eigenvalues is denoted $\sigma_p(A)$ (called point spectrum) and is defined by

$$\sigma_p(A) = \left\{ \lambda \in \sigma(A) : N(A - \lambda I) \neq \{0\} \right\}.$$

3.6.2   The Essential Spectrum of a Bounded Linear Operator

Definition 3.50. The essential spectrum $\sigma_e(A)$ of a bounded linear operator A : $\mathbb{E} \mapsto \mathbb{E}$ is defined by

$$\sigma_e(A) := \left\{ \lambda \in \mathbb{K} : \lambda I - A \text{ is not a Fredholm operator of index } 0 \right\}.$$

If $\lambda \in \mathbb{K}$ does not belong to neither $\sigma_p(A)$ nor $\sigma_e(A)$, then $\lambda I - A$ must be injective $(N(\lambda I - A) = \{0\})$ and $R(\lambda I - A)$ is closed with

$$0 = \dim N(\lambda I - A) = \dim \mathbb{E} \setminus R(\lambda I - A).$$

Consequently, $(\lambda I - A)$ must be bijective which yields $\lambda \in \rho(A)$. In view of the previous facts, we have,

$$\sigma(A) = \sigma_p(A) \cup \sigma_e(A).$$

Remark 3.51. It should be noted that the union $\sigma(A) = \sigma_p(A) \cup \sigma_e(A)$ is not a disjoint one. Indeed, the intersection $\sigma_p(A) \cap \sigma_e(A)$ consists of eigenvalues λ of A for which:

(a) either $\dim N(\lambda I - A)$ is not finite
(b) or $R(\lambda I - A)$ is not closed
(c) or $\dim N(A) \neq \dim \mathbb{E} \setminus R(A)$.

Definition 3.52. The continuous spectrum $\sigma_c(A)$ of a bounded linear operator A : $\mathbb{E} \mapsto \mathbb{E}$ is defined by,

$$\sigma_c(A) := \left\{ \lambda \in \sigma_e(A) \setminus \sigma_p(A) : \overline{R(\lambda I - A)} = \mathbb{E} \right\}.$$

Definition 3.53. The residual spectrum $\sigma_r(A)$ of a bounded linear operator $A : \mathbb{E} \mapsto \mathbb{E}$ is defined by,

$$\sigma_r(A) := \left(\sigma_e(A) \setminus \sigma_p(A) \right) \setminus \sigma_c(A).$$

As in the classical operator theory, we have

$$\sigma(A) = \sigma_p(A) \cup \sigma_c(A) \cup \sigma_r(A).$$

Example 3.54. Consider the diagonal operator $D : \mathbb{E}_\omega \mapsto \mathbb{E}_\omega$ defined for all $u = (u_j)_{j \in \mathbb{N}} \in \mathbb{E}_\omega$ by

$$Du = \sum_{j=0}^{\infty} \lambda_j u_j e_j,$$

where $\lambda = (\lambda_j)_{j \in \mathbb{N}}$ with $\lambda_j \in \mathbb{K}$ for all $j \in \mathbb{N}$ and

$$\sup_{j \in \mathbb{N}} |\lambda_j| < \infty.$$

Proposition 3.55. $\sigma(D) = \overline{\{\lambda_k : k \in \mathbb{N}\}}$ *the closure in* $\mathbb{K}$ *of* $\{\lambda_k : k \in \mathbb{N}\}$. *That is,*

$$\sigma(D) = \left\{ \lambda \in \mathbb{K} : \inf_{j \in \mathbb{N}} |\lambda - \lambda_j| = 0 \right\}.$$

Proof. It is sufficient to prove that $\lambda \in \rho(D)$ if and only if

$$\inf_{j \in \mathbb{N}} |\lambda - \lambda_j| > 0. \tag{3.5}$$

Suppose Eq. (3.5) holds and define the linear operator D' on $\mathbb{E}_\omega$ for all $u = (u_j)_{j \in \mathbb{N}} \in \mathbb{E}_\omega$ by

$$D'u = \sum_{j=0}^{\infty} \frac{u_j e_j}{\lambda - \lambda_j}.$$

It is easy to see that the operator D' is well-defined, bounded, and that its norm is given as follows:

$$\left\| D' \right\| = \sup_{j \in \mathbb{N}} \frac{\left\| D' e_j \right\|}{\left\| e_j \right\|} = \sup_{j \in \mathbb{N}} \left| \frac{1}{\lambda - \lambda_j} \right| = \frac{1}{\inf_{j \in \mathbb{N}} |\lambda - \lambda_j|} < \infty.$$

It is also easy to check that, $(\lambda I - D)D' = D'(\lambda I - D) = I$, that is, $D' = (\lambda I - D)^{-1}$ and therefore $\lambda \in \rho(D)$.

Conversely, suppose $\lambda \in \rho(D)$ with $\inf_{j \in \mathbb{N}} |\lambda - \lambda_j| = 0$. Clearly, there exists $(\lambda_{j_k})_{k \in \mathbb{N}} \subset (\lambda_j)_{j \in \mathbb{N}}$ a subsequence such that

$$\lim_{k \to \infty} \frac{\left\| (\lambda I - D) e_{j_k} \right\|}{\left| \omega_{j_k} \right|^{1/2}} = \lim_{k \to \infty} |\lambda - \lambda_{j_k}| = 0. \tag{3.6}$$

Now using the fact that $(\lambda I - D)D' = D'(\lambda I - D) = I$ it follows that

$$\frac{\left\| (\lambda I - D) e_{j_k} \right\|}{\left| \omega_{j_k} \right|^{1/2}} \geq \frac{1}{\left\| (\lambda I - D)^{-1} \right\|}$$

which yields

$$\lim_{k \to \infty} \frac{\left\| (\lambda I - D)e_{j_k} \right\|}{\left| \omega_{j_k} \right|^{1/2}} \geq \frac{1}{\left\| (\lambda I - D)^{-1} \right\|} > 0. \tag{3.7}$$

From Eqs. (3.6) and (3.7), we obtain a contradiction. Therefore, $\lambda \in \rho(D)$ yields Eq. (3.5).

3.7 Bibliographical Notes

The material of this chapter is mainly taken from the following sources: Diagana [13, 16], Diagana et al. [18], Diarra [19–21], Perez-Garcia [44], Perez-Garcia and Vega [43] Schneider [52], Serre [48], Gohberg et al. [27], and Śliwa [51]. For additional reading upon bounded linear operators and their spectral theory, we refer the reader to Vishik [54] and Davies [11].

Chapter 4
The Vishik Spectral Theorem

This chapter is devoted to the so-called Vishik spectral theorem for bounded linear operators. Here, we mainly follow Attimu [4], Attimu and Diagana [3], Baker [7], and Vishik [54].

4.1 The Shnirel'man Integral and Its Properties

This section is devoted to the study of the so-called Shnirel'man integral [54]. Such an integral was introduced in the literature in 1938. It plays a crucial role in various areas including, but are not limited to, the construction of the Vishik spectral theorem, and in transcendental number theory. Further, it can be utilized to prove a non-archimedean version of the Cauchy integral theorem, the residue theorem, and the maximum principle.

In this chapter, we suppose that our non-archimedean valued field $(\mathbb{K}, |\cdot|)$ is algebraically closed with dense valuation.

4.1.1 Basic Definitions

Let $\sigma \subset \mathbb{K}$ be a subset and let $r > 0$. The sets $D(\sigma, r)$ and $D(\sigma, r^-)$ are defined respectively by,

$$D(\sigma, r) := \left\{ x \in \mathbb{K} : dist(x, \sigma) \leq r \right\}$$

© The Author(s) 2016
T. Diagana, F. Ramaroson, *Non-Archimedean Operator Theory*,
SpringerBriefs in Mathematics, DOI 10.1007/978-3-319-27323-5_4

and

$$D(\sigma, r^-) := \left\{ x \in \mathbb{K} : dist(x, \sigma) < r \right\},$$

where $dist(x, \sigma) = \inf_{y \in \sigma} |x - y|$.

Moreover, for $a \in \mathbb{K}$, we define the sets $D(a, r^-)$ and $D(a, r)$ respectively by

$$D(a, r^-) := \left\{ x \in \mathbb{K} : |x - a| < r \right\}$$

and

$$D(a, r) := \left\{ x \in \mathbb{K} : |x - a| \le r \right\}.$$

Lemma 4.1 ([7]). *Let $\sigma \subset \mathbb{K}$ be a nonempty compact subset. Then for every $s > 0$, there exist $0 < r \in |\mathbb{K}|$ and $a_1, \cdots, a_N \in \sigma$ such that $r < s$ and*

$$D(\sigma, r) = \bigsqcup_{i=1}^{N} D(a_i, r) \ \ and \ \ \sigma \subset \bigsqcup_{i=1}^{N} D(a_i, r^-),$$

where the symbol $\bigsqcup$ stands for disjoint unions.

Proof. Using the fact that σ is compact it follows hat there exist $a_1, \cdots, a_N \in \sigma$ such that

$$\sigma \subset \bigsqcup_{i=1}^{N} D(a_i, s^-).$$

In fact, we can assume that these unions are disjoint. Now, for each $1 \le i \le N$, $D(a_i, s^-)$ is closed, and hence $\sigma \cap D(a_i, s^-)$ is compact. Therefore, there exist $b_1, \cdots, b_N \in \sigma$ such that for $1 \le i \le N$,

$$\left| b_i - a_i \right| = \sup \left\{ \left| x - a_i \right| : x \in \sigma \cap D(a_i, s^-) \right\}.$$

Since each $b_i \in D(a_i, s^-)$, it follows that for $1 \le i \le N$, $s > |b_i - a_i|$. Using the fact that $|K|$ is dense in $\mathbb{R}$, then there exists $r \in |\mathbb{K}|$ such that $r > 0$ and

$$s > r > \left| b_1 - a_1 \right|, \cdots, \left| b_N - a_N \right|.$$

We now show that

$$\sigma \subset \bigsqcup_{i=1}^{N} D(a_i, r^-).$$

Indeed, let $a \in \sigma \subset \bigsqcup_{i=1}^{N} D(a_i, s^-)$, then for some $1 \leq i \leq N$, $a \in D(a_i, s^-)$ and hence

$$\left| a - a_i \right| \leq \sup \left\{ \left| x - a \right| : x \in \sigma \cap D(a_i, s^-) \right\} = \left| b_i - a_i \right| < r.$$

Consequently, $a \in D(a_i, r^-)$ and $\sigma \subset \bigsqcup_{i=1}^{N} D(a_i, r^-)$.

Obviously, $D(a_i, r)$'s are disjoint as the $D(a_i, s^-)$'s are disjoint and $D(a_i, r) \subset D(a_i, s^-)$ for all $1 \leq i \leq N$. It remains to show that

$$D(\sigma, r) = \bigsqcup_{i=1}^{N} D(a_i, r).$$

Moreover, we have

$$\bigsqcup_{i=1}^{N} D(a_i, r) \subset D(\sigma, r).$$

To prove the other inclusion, let $x \in D(\sigma, r)$. Using the fact that σ is compact it follows that there exists $a \in \sigma$ such that $r \geq \text{dist}(x, \sigma) = |x - a|$. Hence, $x \in D(a, r)$. From the first part of the lemma, it follows that there exists $1 \leq j \leq N$ such that $a \in D(a_j, r^-) \subset D(a_j, r)$. Hence, $D(a, r) = D(a_j, r)$, which shows that

$$D(\sigma, r) \subset \bigsqcup_{i=1}^{N} D(a_i, r).$$

Observe that the result in Baker [7, Lemma 1.3] can be generalized as follows:

Lemma 4.2 ([3]). *Let $\emptyset \neq \sigma \subset \mathbb{K}$ and let $r > 0$. Then if I is a nonempty set and if $\{b_i : i \in I\} \subset \mathbb{K}$ is a subset such that*

$$\sigma \subset \bigsqcup_{i \in I} D(b_i, r^-),$$

then there exist a subset $J \subset I$ and subset $\{a_j : j \in J\} \subset \sigma$ such that

$$D(\sigma, r^-) = \bigsqcup_{j \in J} D(a_j, r^-) = \bigsqcup_{j \in L} D(b_j, r^-), \quad and$$

$$D(\sigma, r) = \bigsqcup_{j \in L} D(a_j, r) = \bigsqcup_{j \in J} D(b_j, r).$$

Proof. Let $J := \{j \in I : D(b_j, r^-) \cap \sigma \neq \emptyset\}$ and rewrite J as $J = \{i_j : j \in J\}$. For all $j \in J$, choose $a_j \in D(b_{i_j}, r^-)$. Now, $D(b_{i_j}, r^-) = D(a_j, r^-)$, and therefore,

$$\sigma \subset \bigsqcup_{j \in J} D(a_j, r^-) = \bigsqcup_{j \in J} D(b_{i_j}, r^-) = \bigsqcup_{j \in J} D(b_j, r^-).$$

Clearly,

(i) $\bigsqcup_{j \in J} D(a_j, r^-) \subset D(\sigma, r^-)$;

(ii) $\bigsqcup_{j \in J} D(a_j, r) \subset D(\sigma, r)$.

Now we show the reverse inclusions. Let $x \in D(\sigma, r^-)$. Then $dist(x, \sigma) < r$ and hence, there exists $a \in \sigma$ such that $\left|x - a\right| < r$. Since $a \in \sigma = \bigsqcup_{j \in J} D(a_j, r^-)$, there exists $a_{j_0} \in \sigma$ with $j_0 \in J$ such that $\left|a - a_{j_0}\right| < r$.

Now

$$\left|x - a_{j_0}\right| \leq \max\{\left|x - a\right|, \left|a - a_{j_0}\right|\} < r,$$

and hence $x \in \bigsqcup_{j \in J} D(a_j, r^-)$. Therefore, $D(\sigma, r^-) \subset \bigsqcup_{j \in J} D(a_j, r^-)$.

Finally, let $x \in D(\sigma, r)$, that is, $dist(x, \sigma) \leq r$ and there exists $a \in \sigma$ such that $\left|x - a\right| \leq r$. Again, since $a \in \sigma$, there exists $a_{j_1} \in \sigma$ such that $\left|a - a_{j_1}\right| < r$.

Now

$$\left|x - a_{j_1}\right| \leq \max\left\{\left|x - a\right|, \left|a - a_{j_1}\right|\right\} \leq r,$$

and hence $x \in \bigsqcup_{j \in J} D(a_j, r)$ and therefore,

$$D(\sigma, r) \subset \bigsqcup_{j \in J} D(a_j, r).$$

Corollary 4.3 ([3]). *Let $\emptyset \neq \sigma \subset \mathbb{K}$ and let $r > 0$. Let $b_1, \cdots, b_M$ be in $\mathbb{K}$ with*

$$\sigma \subset \bigsqcup_{i=1}^{M} D(b_i, r^-).$$

Then there exist $a_1, \cdots, a_N$ *in* σ *and* $\emptyset \neq J \subset \{1, \cdots, M\}$ *such that the* $D(a_i, r)$ *are disjoint and*

$$D(\sigma, r^-) = \bigsqcup_{i=1}^{N} D(a_i, r^-) = \bigsqcup_{i \in J} D(b_i, r^-) \ \text{and} \ D(\sigma, r) = \bigsqcup_{i=1}^{N} D(a_i, r) = \bigsqcup_{i \in J} D(b_i, r).$$

$$(4.1)$$

Proof. The proof follows from Lemma 4.2. Take $I = \{1, \cdots, M\}$ and the existence of the set $J \subset I$ such that Eq. (4.1) is satisfied is guaranteed and the proof is complete.

The concept of (local) analyticity in the next definition plays a crucial role in this section.

Definition 4.4 ([3]). Let $a \in \mathbb{K}$ and let $r > 0$. A function $f : D(a, r) \mapsto \mathbb{K}$ is said to be analytic if f can be represented by a power series on $D(a, r)$, that is,

$$f(x) = \sum_{k=0}^{\infty} c_k(x - a)^k \ \text{with} \ \lim_{k \to \infty} r^k |c_k| = 0.$$

Definition 4.5. Let $a \in \mathbb{K}$ and let $r > 0$. The function $f : D(a, r) \mapsto \mathbb{K}$ is said to be "Krasner analytic" if it is a uniform limit of rational functions with poles belong to the complement of $D(a, r)$.

It should be mentioned that if $r \in |\mathbb{K}|$, it can be shown that a function that is analytic over $D(a, r)$ in the sense of Krasner is also analytic in the sense of Definition 4.4.

Definition 4.6 ([3]). Let $\emptyset \neq \sigma \subset \mathbb{K}$ and let $r > 0$. Let $B_r(\sigma)$ denote the collection of all functions $f : D(\sigma, r) \to \mathbb{K}$ which are analytic on $D(a, r)$ for $a \in \mathbb{K}$ and $D(a, r) \subset D(\sigma, r)$.
If f is bounded on $D(\sigma, r)$, then we set

$$\left\| f \right\|_r := \max_{x \in D(\sigma, r)} \left| f(x) \right|.$$

It should be indicated that the notion of local analyticity in Definition 4.7 is due to Baker [7].

Definition 4.7 ([3]). Let $\emptyset \neq \sigma \subset \mathbb{K}$. Define $L(\sigma)$ to be the collection of all $\mathbb{K}$-valued functions f for which there exist $a_1, \cdots, a_N$ in $\mathbb{K}$ and $0 < r \in |\mathbb{K}|$ such that

$$\sigma \subset \bigsqcup_{i=1}^{N} D(a_i, r^-),$$

where the $D(a_i, r)$ are disjoint and f is analytic on each $D(a_i, r)$. The collection of functions $L(\sigma)$ will be called the set of locally analytic functions on σ.

From Definition 4.7 we deduce that $\mathrm{Dom}(f)$, the domain of $f \in L(\sigma)$ satisfies,

$$\mathrm{Dom}(f) \subset \bigsqcup_{i=1}^{N} D(a_i, r).$$

Further, $L(\sigma) \neq \emptyset$ since polynomials belong to it.

Theorem 4.8 ([7]). *Let $\emptyset \neq \sigma \subset \mathbb{K}$ be a compact subset. Then*

$$L(\sigma) = \bigcup_{r>0} B_r(\sigma). \tag{4.2}$$

Proof. Let $f \in L(\sigma)$. Then by Definition 4.7, there exist $a_1, \cdots, a_N$ in $\mathbb{K}$ and $0 < s \in |\mathbb{K}|$ such that

$$\sigma \subset \bigsqcup_{i=1}^{N} D(a_i, s^-), \tag{4.3}$$

where the $D(a_i, s)$ are disjoint and f is Krasner analytic on each $D(a_i, s)$.

Using Corollary 4.3, it follows that there exists $\emptyset \neq J \subset \{1, \cdots, N\}$ such that

$$D(\sigma, s) = \bigsqcup_{j \in J} D(a_j, s). \tag{4.4}$$

Let $b \in \mathbb{K}$, with $D(b, s) \subset D(\sigma, s)$. Then from Eq. (4.4), there exists $j \in J$ such that $b \in D(a_j, s)$ and hence $D(a_j, s) = D(b, s)$. Since f is Krasner analytic on $D(a_j, s)$, and hence on $D(b, s)$, it follows that $f \in B_s(\sigma)$. Consequently,

$$L(\sigma) \subset \bigcup_{r>0} B_r(\sigma). \tag{4.5}$$

To prove the other inclusion, let $f \in \bigcup_{r>0} B_r(\sigma)$. Hence there exists $s > 0$ such that $f \in B_s(\sigma)$. From the compactness of σ and Corollary 4.3 there exist $b_1, \cdots, b_M \in \mathbb{K}$ such that

$$\sigma \subset D_\sigma(s) = \bigsqcup_{i=1}^{N} D(b_i, s). \tag{4.6}$$

Now from Lemma 4.1, there exist $a_1, \cdots, a_N$ in σ and $0 < r \in |\mathbb{K}|$ such that $r < s$ and

$$D(\sigma, r) = \bigsqcup_{i=1}^{N} D(a_i, r), \quad \sigma \subset \bigsqcup_{i=1}^{N} D(a_i, r^-).$$

Let $1 \leq i \leq N$ be arbitrary. Then since $a_i \in \sigma$, Eq. (4.6) yields there exists $1 \leq j \leq M$ such that $a_i \in D(b_j, s)$ and hence $D(a_i, s) = D(b_j, s)$. But $f \in B_s(\sigma)$ and hence f is Krasner analytic on $D(b_j, s) = D(a_i, s)$. It follows that f is Krasner analytic on $D(a_i, r) \subset D(a_i, s)$, since $r < s$. Therefore,

$$\bigcup_{r>0} B_r(\sigma) \subset L(\sigma). \tag{4.7}$$

Now we obtain the result using Eqs. (4.5) and (4.7).

Remark 4.9. Let us mention that the notion of local analyticity in Definition 4.7 generalizes that of Koblitz [32, p. 136], in which the local analyticity on compact $\emptyset \neq \sigma \subset \mathbb{K}$ was defined as

$$L(\sigma) = \bigcup_{r>0} B_r(\sigma).$$

4.1.2   The Shnirel'man Integral

Definition 4.10 (Shnirel'man Integral). Let κ be the residue field of $\mathbb{K}$ and let $f(x)$ be a $\mathbb{K}$-valued function defined for all $x \in \mathbb{K}$ such that $|x - a| = r$ where $a \in \mathbb{K}$ and $r > 0$ with $r \in |\mathbb{K}|$. Let $\Gamma \in \mathbb{K}$ be such that $|\Gamma| = r$. Then the Shnirel'man integral of f is defined by

$$\int_{a,\Gamma} f(x)dx := \lim_{n \to \infty} {}' \frac{1}{n} \sum_{\eta^n = 1} f(a + \eta\Gamma), \tag{4.8}$$

if the limit exists.

The prime on the right hand side of Eq. (4.8) means that if $char(\kappa) = p$, then the additional condition $gcd(p, n) = 1$ is imposed.

Lemma 4.11 ([3]).

(i) *Suppose that f is bounded on the circle $|x - a| = r$. If $\int_{a,\Gamma} f(x)dx$ exists, then*

$$\left| \int_{a,\Gamma} f(x)dx \right| \leq \max_{|x-a|=r} \left| f(x) \right|.$$

(ii) *The integral $\int_{a,\Gamma}$ commutes with limits of functions which are uniform limits*
on $\{x \in \mathbb{K} : |x - a| = r\}$.

(iii) *If $r_1 \leq r \leq r_2$ and $f(x)$ is given by a convergent Laurent series $\sum_{k \in \mathbb{Z}} c_k(x - a)^k$*
in the annulus $r_1 \leq |x - a| \leq r_2$, then

$$\int_{a,\Gamma} f(x)dx = c_0$$

and is independent of the choice of Γ with $|\Gamma| = r$, as long as $r_1 \leq r \leq r_2$.
More generally,

$$\int_{a,\Gamma} \frac{f(x)}{(x - a)^k} dx = c_k.$$

Proof. The proof of statements (i) and (ii) follow directly from the definition of the
Shnirel'man integral. To prove (iii), note that for $k \neq 0$ and $n > |k|$,

$$\sum_{\eta^n = 1} \eta^k = 0, \tag{4.9}$$

and hence

$$f(a + \Gamma\eta) = c_0 + \sum_{k \in \mathbb{Z} \setminus \{0\}} c_k \Gamma^k \eta^k.$$

The result is now a consequence of Eq. (4.9), Eq. (4.8) and the fact that
$\lim_{k \to \infty} c_k \Gamma^k = 0$.

Lemma 4.12 ([3]). *Fix $x_0 \in \mathbb{K}$ and $m > 0$. Then, the following holds*

$$\int_{a,\Gamma} \frac{dx}{(x - x_0)^m} = \begin{cases} 0 & \text{if } |a - x_0| < r; \\ \\ (a - x_0)^{-m} & \text{if } |a - x_0| > r. \end{cases}$$

Proof. The result follows from the fact that for $|x - a| = r$, we have the following
Laurent expansion

$$\frac{1}{(x - x_0)^m} = \begin{cases} \left(\sum_{k \geq 0} (x_0 - a)^k (x - a)^{-k-1} \right)^m & \text{if } |a - x_0| < r; \\ \\ \left(\frac{1}{a - x_0} \sum_{k \geq 0} (x_0 - a)^{-k} (x - a)^k \right)^m & \text{if } |a - x_0| > r. \end{cases} \tag{4.10}$$

To obtain the desired results, one makes use of (iii) of Lemma 4.11 with $r = r_1 = r_2$.

Corollary 4.13. *Fix $x_0 \in \mathbb{K}$ and $m > 1$. Then, the following hold*

$$\int_{a,\Gamma} \frac{x - a}{(x - x_0)} \, dx = \begin{cases} 1 & \text{if } \left| a - x_0 \right| < r; \\ 0 & \text{if } \left| a - x_0 \right| > r. \end{cases}$$

and

$$\int_{a,\Gamma} \frac{x - a}{(x - x_0)^m} \, dx = \begin{cases} 0 & \text{if } \left| a - x_0 \right| < r; \\ 0 & \text{if } \left| a - x_0 \right| > r. \end{cases}$$

Lemma 4.14 (Nonarchimedean Cauchy Integral Formula).   *If f is analytic on $D(a, r)$ and if $|\Gamma| = r \in |\mathbb{K}|$, then*

$$\int_{a,\Gamma} \frac{f(x)(x - a)}{(x - x_0)} \, dx = \begin{cases} f(x_0) & \text{if }, \left| a - x_0 \right| < r, \\ 0 & \text{if } \left| a - x_0 \right| > r. \end{cases} \tag{4.11}$$

In particular, this integral does not depend on the choice of a, Γ or r as long as $|x_0 - x|$ is either less than or greater than r.

Proof. To prove Eq. (4.11), we first assume that $|x_0 - a| < r$ and write

$$\int_{a,\Gamma} \frac{f(x)(x - a)}{(x - x_0)} \, dx = \int_{a,\Gamma} \frac{(f(x) - f(x_0))(x - a)}{(x - x_0)} \, dx + f(x_0) \int_{a,\Gamma} \frac{(x - a)}{(x - x_0)} \, dx. \tag{4.12}$$

Then, the second integral on the right hand side of Eq. (4.12) equals 1 by suing Corollary 4.13. We show that the first integral is 0. Note that

$$f(x) - f(x_0) = \sum_{k \geq 1} c_k \left((x - a)^k - (x_0 - a)^k \right)$$

$$= (x - x_0) \sum_{k \geq 1} c_k \left[\sum_{i=0}^{k-1} (x - a)^{k-1-i} (x_0 - a)^i \right].$$

Now

$$\int_{a,\Gamma} \frac{(f(x) - f(x_0))(x-a)}{(x-x_0)} \, dx = \int_{a,\Gamma} \sum_{k\geq1} c_k \left[\sum_{i=0}^{k-1} (x-a)^{k-i}(x_0-a)^i \right] dx$$

$$= \sum_{k\geq1} c_k \int_{a,\Gamma} \left[\sum_{i=0}^{k-1} (x-a)^{k-i}(x_0-a)^i \right] dx$$

$$= 0,$$

by using (ii)–(iii) of Lemma 4.11.

Now if $|x_0 - a| > r$ then we can apply Eq. (4.10) and (iii) of Lemma 4.11 to obtain

$$\int_{a,\Gamma} \frac{f(x)(x-a)}{(x-x_0)} \, dx = 0.$$

Theorem 4.15 (Non-archimedean Residue Theorem [54]).   *Let f be a rational function over $\mathbb{K}$ and suppose none of the poles x_0 of f satisfy $|x_0 - a| = |\Gamma|$, where $\Gamma \in \mathbb{K} - \{0\}$. Then*

$$\int_{a,\Gamma} f(x)(x-a) \, dx = \sum_{|x_0-a|<|\Gamma|} res_{x=x_0} f(x), \tag{4.13}$$

where $res_{x=x_0} f(x)$ is the coefficient of $(x - x_0)^{-1}$ in the Laurent expansion of f about x_0.

Proof. We expand f in partial fractions and make use of (iii) of Lemma 4.11 and Corollary 4.13.

Theorem 4.16.   *Let f be analytic on $D(a, r)$. Then for any $x_0 \in \mathbb{K}$, we have*

$$\int_{a,\Gamma} \frac{f(x)}{(x-x_0)} \, dx = \begin{cases} \dfrac{f(a)-f(x_0)}{a-x_0} & \text{if } |a - x_0| < r; \\[2ex] \dfrac{f(a)}{a-x_0} & \text{if } |a - x_0| > r. \end{cases}$$

Proof. Let $|x_0 - a| < r$. Then from Eq. (4.10) with $m = 1$, we have

$$\frac{f(x)}{x-x_0} = \sum_{i\geq0} \frac{f^{(i)}(a)}{i!}(x-a)^i \cdot \sum_{k\geq0} (x_0-a)^{k-1}(x-a)^{-k}$$

$$= \sum_{i,j-1\geq0} \frac{f^{(i)}(a)}{i!}(x-a)^{i-j}(x_0-a)^{j-1}.$$

Hence using (iii) of Lemma 4.11, we have

$$\int_{a,\Gamma} \frac{f(x)}{x - x_0} dx = \sum_{j \geq 1} \frac{f^{(j)}(a)}{j!} (x_0 - a)^{j-1}$$

$$= \frac{1}{x_0 - a} \sum_{j \geq 1} \frac{f^{(j)}(a)}{j!} (x_0 - a)^j$$

$$= \frac{f(x_0) - f(a)}{x_0 - a}.$$

Clearly, if $|x_0 - a| > r$, then

$$\frac{f(x)}{x - x_0} = (-1) \sum_{i \geq 0} \frac{f^{(i)}(a)}{i!} (x - a)^i \cdot \sum_{j \geq 0} (x_0 - a)^{-j-1} (x - a),$$

and hence

$$\int_{a,\Gamma} \frac{f(x)}{x - x_0} dx = \frac{f(a)}{a - x_0}.$$

Remark 4.17. Using Theorem 4.16 and [(iii), Lemma 4.11] it follows that if f is analytic on $D(a, r)$ and $x_0 \in D(a, r^-)$, then

$$(a - x_0) \int_{a,\Gamma} \frac{f(x)}{(x - x_0)} dx + \int_{x_0,\Gamma} f(x) dx = \int_{a,\Gamma} f(x) dx.$$

Moreover, if $x_0 \notin D(a, r)$, then

$$\int_{a,\Gamma} f(x) dx = (a - x_0) \int_{a,\Gamma} \frac{f(x)}{x - x_0} dx.$$

Corollary 4.18. *If f is analytic on $D(a, r)$, then*

$$\int_{a,\Gamma} f(x) dx = \begin{cases} (a - x_0) \displaystyle\int_{a,\Gamma} \frac{f(x)}{x - x_0} dx + \int_{x_0,\Gamma} f(x) dx & \text{if } |a - x_0| < r, \\[4mm] (a - x_0) \displaystyle\int_{a,\Gamma} \frac{f(x)}{x - x_0} dx & \text{if } |a - x_0| > r. \end{cases}$$

4.2 Distributions with Compact Support

Let $\sigma \subset \mathbb{K}$ be a compact subset and let $r > 0$. It follows that there is a finite set I with $a_i \in \mathbb{K}$ for $i \in I$ with

$$\sigma \subset \bigsqcup_{i \in I} D(a_i, r^-), \quad \text{and} \quad D(\sigma, r^-) = \bigsqcup_{i \in I} D(a_i, r^-). \tag{4.14}$$

Let $f \in B_r(\sigma)$. Then $f : D(\sigma, r^-) \to \mathbb{K}$ is clearly analytic and hence satisfies:

(i) $f(x) = \sum_{j \in \mathbb{N}} f_{ij}(x - a_i)^j$ for $x \in D(a_i, r^-)$.

(ii) For all $i \in I$, $|f_{ij}| r^j \to 0$ as $j \to \infty$.

(iii) The norm of f is defined as

$$\left\| f \right\|_r := \sup_{i \in I, j \in \mathbb{N}} |f_{ij}| r^j.$$

Let us mention that $(B_r(\sigma), \| \cdot \|_r)$ is a non-archimedean Banach space. Furthermore, the following embedding is continuous

$$B_r(\sigma) \hookrightarrow B_{r_1}(\sigma)$$

with $0 < r_1 < r$.

Recall that (from Theorem 4.8) the following holds

$$L(\sigma) = \bigcup_{r > 0} B_r(\sigma).$$

Definition 4.19 ([3]). The space $L^*(\sigma) := L(\sigma)^*$ (topological dual of $L(\sigma)$) is called the space of distributions with support σ.

For all $\mu \in L^*(\sigma)$ and $f \in L(\sigma)$, we represent the canonical pairing between μ and f as

$$(\mu, f) = (\mu(x), f(x)) = \mu(f).$$

Moreover, it is easy to see that for $\mu \in L^*(\sigma)$, then $\mu_{|B_r(\sigma)}$ is a continuous linear functional whose norm is

$$\left\| \left\| \mu \right\| \right\|_r := \sup_{f \in B_r(\sigma), f \neq 0} \frac{|\mu(f)|}{\left\| f \right\|_r}.$$

In particular, if $0 < r_1 < r$, then

$$\Big| \big\| \mu \big\| \Big|_{r_1} \geq \Big| \big\| \mu \big\| \Big|_r.$$

For $r > 0,\ i \in I, j \in \mathbb{N}$ and $x \in \mathbb{K}$, we define

$$\chi(r, i, j; x) = \begin{cases} (x - a_i)^j & \text{if } |x - a_i| < r, \\[2mm] 0 & \text{if } |x - a_i| \geq 0. \end{cases}$$

Obviously, $\chi(r, i, j; \cdot) \in B_r(\sigma)$.

It can also be shown (see [32]) that the weak topology on $L^*(\sigma)$ whose basis is the neighborhoods of zero given by

$$U_{f,\varepsilon} := \Big\{ \mu \in L^*(\sigma) : \big| \mu(f) \big| < \varepsilon \Big\}$$

and the stronger topology on $L^*(\sigma)$ whose basis is the neighborhoods of zero given by

$$U(r, \varepsilon) := \Big\{ \mu \in L^*(\sigma) : \big\| \mu \big\|_r < \varepsilon \Big\}$$

have the same convergent sequences.

Here, we set $\overline{\sigma} := \mathbb{K} - \sigma$ and

$$\overline{D}(\sigma, r) = \mathbb{K} - D(\sigma, r^-) = \Big\{ x \in \mathbb{K} : \quad dist(x, \sigma) \geq r \Big\}.$$

Definition 4.20. The collection of all functions $\varphi : \overline{\sigma} \to \mathbb{K}$ which are Krasner analytic and vanish at infinity that is:

(i) φ is a limit of rational functions whose poles are contained in σ, the limit being uniform in any set of the form $\overline{D}(\sigma, r)$;
(ii) $\lim\limits_{|z| \to \infty} \varphi(z) = 0$;

is denoted $H_0(\overline{\sigma})$.

For $\varphi \in H_0(\overline{\sigma})$, we define

$$\big\| \phi \big\|_r := \max_{z \in \overline{D}(\sigma, r)} \big| \phi(z) \big| = \max_{dist(z, \sigma) = r} \big| \phi(z) \big|.$$

In particular, for $0 < r_1 < r$, then

$$\left\|\phi\right\|_r \leq \left\|\phi\right\|_{r_1}.$$

As a topology on $H_0(\overline{\sigma})$, we take as a basis the open neighborhoods of zero given by

$$U_0(r, \varepsilon) = \left\{\phi : \|\phi\|_r < \varepsilon\right\}.$$

4.3   Cauchy–Stieltjes and Vishik Transforms

Definition 4.21 (Cauchy–Stieltjes Transform). Let $\sigma \subset \mathbb{K}$ be a compact subset and let $\mu \in L^*(\sigma)$. The Cauchy–Stieltjes transform of μ is the function

$$\varphi = S\mu : \overline{\sigma} \to \mathbb{K}$$

$$z \mapsto \left(\mu(x), \frac{1}{z - x}\right).$$

Let $f \in B_r(\sigma)$ and suppose $\sigma \subset \bigsqcup_{i \in I} D(a_i, r^-)$ where I is a finite index set. Fix $\Gamma \in \mathbb{K}$ such that for all $i \in I$,

$$\sup_{b \in D(a_i, r^-) \cap \sigma} \left|a_i - b\right| < \left|\Gamma\right| < r. \tag{4.15}$$

Definition 4.22 ([54]). We define the Vishik transform V (under the assumptions leading to Eq. (4.15)) by

$$V\varphi : B_r(\sigma) \to \mathbb{K}$$

$$f \mapsto \sum_{i \in I} \int_{a_i, \Gamma} (z - a_i)\varphi(z)f(z)dz.$$

Lemma 4.23 ([54]). *Let $\mu \in L^*(\sigma)$ be a distribution with compact support. Then $S\mu \in H_0(\overline{\sigma})$ and $S : L^*(\sigma) \to H_0(\overline{\sigma})$ is continuous.*

Proof. We first check that $S\mu_{|\overline{D}(\sigma, r)}$ is an analytic element. Let $r_1 < r$ and choose I', a'_k, for all $k \in I'$ as in Eq. (4.14).

Now

$$\phi(z) = S\mu(z)$$

$$= \left(\mu(x), \frac{1}{z-x} \right)$$

$$= \sum_{k \in I'} \left(\mu(x), \chi(r_1, k, 0; x) \frac{1}{z-x} \right)$$

$$= \sum_{k \in I'} \left(\mu(x), \chi(r_1, k, 0; x) \frac{1}{(z-a'_k) - (x-a'_k)} \right).$$

Since $z \in \overline{D}(\sigma, r)$, then

$$\chi(r_1, k, 0; x) \frac{1}{(z-a'_k) - (x-a'_k)} = \sum_{j \in \mathbb{N}} (z-a'_k)^{-j-1} \chi(r_1, k, j; x),$$

and we can approximate $\phi(z) = S\mu(z)$ uniformly on $\overline{D}(\sigma, r)$ by the following rational functions

$$\phi_N(z) = \sum_{k \in I'} \sum_{j=0}^{N} (z-a'_k)^{-j-1} \left(\mu(x), \chi(r_1, k, j; x) \right).$$

In view of the above one can see that $\phi(z) \to 0$ as $|z| \to \infty$ for all N and hence $\lim\limits_{|z| \to \infty} \phi(z) = 0$.

To prove the continuity of $S : L'(\sigma) \to H_0(\overline{\sigma})$, let $\mu \in U(r, \epsilon)$, that is $||\mu||_r < \epsilon$. Then

$$||S\mu||_r = \sup_{z \in \overline{D}(\sigma, r)} |S\mu(z)|$$

$$\leq \sup_{z \in \overline{D}(\sigma, r), N \in \mathbb{N}} |\phi_N(z)|$$

$$\leq \sup_{j \in \mathbb{N}} \epsilon r_1^j r^{-j-1} \leq \epsilon r^{-1},$$

and hence $S\mu \in U_0(r, \epsilon/r)$.

Lemma 4.24 ([54]). *Let* $\varphi \in H_0(\overline{\sigma})$. *Then,*

$$V\varphi(f) := \sum_{i \in I} \int_{a_i, \Gamma} (z-a_i)\varphi(z)f(z)dz, \quad f \in B_r(\sigma),$$

does not depend on the choice of a_i and Γ satisfying (4.15). Furthermore, it is compatible with the inclusion

$$B_r(\sigma) \hookrightarrow B_{r_1}(\sigma) \ \ for \ r_1 < r.$$

In addition, both $V\varphi : B_r(\sigma) \to \mathbb{K}$ and $V : H_0(\overline{\sigma}) \to L^*(\sigma)$ are continuous.

Proof. Suppose for all $i \in I$, $a_i' \in D(a_i, r^-)$, $\Gamma' \in \mathbb{K}$ satisfy Eq. (4.15). Further, suppose $|\Gamma| \le |\Gamma'|$ and that $\phi(z)$ can be approximated uniformly on $\overline{D}(\sigma, |\Gamma|)$ with poles in $D(\sigma, |\Gamma|)$.

Now for all $f \in B_r(\sigma)$,

$$\sum_{i \in I} \int_{a_i, \Gamma} (z - a_i)\phi(z)f(z)dz = \lim_{N \to \infty} \sum_{i \in I} \int_{a_i, \Gamma} (z - a_i)\phi_N(z)f(z)dz$$

$$= \lim_{N \to \infty} \sum_{\rho \in D(\sigma, |\Gamma|^-)} res_{z=\rho}(\phi_N(z)f(z))dz$$

$$= \lim_{N \to \infty} \sum_{i \in I} \int_{a_i', \Gamma'} (z - a_i')\phi_N(z)f(z)dz$$

$$= \sum_{i \in I} \int_{a_i', \Gamma'} (z - a_i')\phi(z)f(z)dz.$$

To show compatibility with $B_r(\sigma) \hookrightarrow B_{r_1}(\sigma)$ where $r_1 < r$, choose I', $a_k' \in \mathbb{K}$ for all $k \in I'$ according to Eq. (4.14) and let Γ' satisfy Eq. (4.15) with r replaced with r_1 and i by k. From the first part, we may assume that $r_1 < |\Gamma| < r$. For all $f \in B_r(\sigma)$,

$$\sum_{i \in I} \int_{a_i, \Gamma} (z - a_i)\phi(z)f(z)dz = \lim_{N \to \infty} \sum_{i \in I} \int_{a_i, \Gamma} (z - a_i)\phi_N(z)f(z)dz,$$

where ϕ_N are rational functions with poles in $D(\sigma, |\Gamma'|^-)$ approximating ϕ uniformly on $\overline{D}(\sigma, |\Gamma'|)$.

Now

$$\sum_{i \in I} \int_{a_i, \Gamma} (z - a_i)\phi_N(z)f(z)dz = \sum_{\rho \in D(\sigma, |\Gamma'|^-)} res_{z=\rho}(\phi_N(z)f(z))$$

$$= \sum_{k \in I'} \int_{a_k', \Gamma'} (z - a_k')\phi_N(z)f(z)dz.$$

In the limit,

$$V\phi(f) = V\phi(f_{|B_{r_1}(\sigma)}), \ \ f \in B_r(\sigma).$$

To prove the continuity of $V\phi : B_r(\sigma) \to \mathbb{K}$, note that for $f \in B_r(\sigma)$,

$$|(V\phi, f)| = |V\phi(f)|$$

$$= \left| \sum_{i \in I} \int_{a_i, \Gamma} (z - a_i)\phi(z)f(z)dz \right|$$

$$\leq r \sup_{|z-a_i|=|\Gamma|} |\phi(z)| \|f\|_r.$$

Hence,

$$\|V\phi\|_r \leq r \sup_{i \in I, |z-a_i|=|\Gamma|} |\phi(z)| < \infty,$$

by using the fact that $\phi \in H_0(\overline{\sigma})$.

Finally, we show that $V : H_0(\overline{\sigma}) \to L'(\sigma)$ is continuous. Indeed if $\phi \in U_0(r, \epsilon)$, that is $\|\phi\|_r < \epsilon$, then

$$\|V\phi\|_r \leq r\epsilon \text{ yields } V\phi \in U(r, r\epsilon).$$

Lemma 4.25 ([54]). $VS = SV = I$.

Proof. Let $\mu \in L'(\sigma)$, $f \in B_r(\sigma)$ with I and Γ chosen as in Eqs. (4.14) and (4.15). Now

$$(VS\mu, f) = \sum_{i \in I} \int_{a_i, \Gamma} (z - a_i)S\mu(z)f(z)dz$$

$$= \sum_{i \in I} \int_{a_i, \Gamma} (z - a_i) \left(\mu(x), \frac{1}{z - x} \right) f(z)dz. \tag{4.16}$$

Now let $r_1 < |\Gamma|$ be such that for all $i \in I$,

$$\sup_{b \in D(a,r^-) \cap \sigma} |b - a_i| < r_1. \tag{4.17}$$

From Eq. (4.16), Eq. (4.17) and (ii) of Lemma 4.11, we obtain

$$(VS\mu, f) = \left(\mu(x), \sum_{i \in I} \int_{a_i, \Gamma} \frac{z - a_i}{z - x} f(z)dz \right)$$

$$= (\mu(x), f(x)) = (\mu, f),$$

by using Theorem 4.14 and hence $VS = Id$.

Let $\phi \in H_0(\overline{\sigma})$. It suffices to show that $SV\phi(z) = \phi(z)$ for $z \in \overline{\sigma}$. Let $a \in \sigma$ and choose Γ so that

$$\sup_{x \in \sigma} |x - a| < |\Gamma|.$$

Now, suppose $|z - a| > |\Gamma|$. Then,

$$S(V\phi)(z) = \left(V\phi(x), \frac{1}{z - x} \right) \quad \text{for } z \in \overline{\sigma},$$

$$= \int_{a,\Gamma} (x - a)\phi(x) \frac{1}{z - x} dx$$

$$= \int_{a,\Gamma} (x - a)\phi(x) \sum_{i \in \mathbb{N}} (x - a)^i (z - a)^{-i-1} dx$$

$$= \int_{a,\Gamma} \phi(x) \sum_{i \geq 1} (x - a)^i (z - a)^{-i} dx.$$

Next we expand ϕ in Laurent series as follows

$$\phi(x) = \sum_{j \geq 1} \phi_j (x - a)^{-j},$$

with the zeroth term absent as $\lim_{|z| \to \infty} \phi(z) = 0$ as $\phi \in H_0(\overline{\sigma})$.

Now, for $|x - a| = |\Gamma|$, the series converges uniformly and

$$S(V\phi)(z) = \sum_{i,j \geq 1} \phi_j (x - a)^{i-j} (z - a)^{-i}$$

$$= \sum_{i,j \geq 1} \phi_j \delta_j^i (z - a)^{-i}$$

$$= \sum_{j \geq 1} \phi_j (z - a)^{-j} = \phi(z).$$

4.4   Analytic Bounded Linear Operators

Definition 4.26. An operator $A \in B(\mathbb{X})$ is called analytic with compact spectrum if $\sigma(A) \subset \mathbb{K}$ is compact, and for all $h \in \mathbb{X}^*$ and $u \in \mathbb{X}$, the function defined by

$$z \mapsto \langle h, R_A(z)u \rangle$$

belongs to $H_0(\overline{\sigma(A)})$.

Classical examples of analytic linear operators with compact support include completely continuous linear operators on a non-archimedean Banach space $\mathbb{X}$.

Example 4.27. Let $p \geq 2$ be a prime and let $\mathbb{K} = \mathbb{Q}_p$ equipped with its p-adic topology. Let $C(\mathbb{Z}_p, \mathbb{Q}_p)$ be the Banach space of all continuous functions from $\mathbb{Z}_p = \left\{ z \in \mathbb{Q}_p : |z| \leq 1 \right\}$ into $\mathbb{Q}_p$ equipped with the sup norm defined by $\|\varphi\|_\infty = \sup_{z \in \mathbb{Z}_p} |\varphi(z)|$ for each $\varphi \in C(\mathbb{Z}_p, \mathbb{Q}_p)$.

Consider the (bounded) position operator, $A : C(\mathbb{Z}_p, \mathbb{Q}_p) \to C(\mathbb{Z}_p, \mathbb{Q}_p)$ defined by

$$A\varphi(x) = x\varphi(x)$$

for all $\varphi \in C(\mathbb{Z}_p, \mathbb{Q}_p)$.

It can be shown that the spectrum $\sigma_A = \mathbb{Z}_p$ and that for all $\xi \in C(\mathbb{Z}_p, \mathbb{Q}_p)^*$ and $u \in C(\mathbb{Z}_p, \mathbb{Q}_p)$, the function $z \mapsto \langle \xi, R_A(z)u \rangle$ belongs to $H_0(\overline{\mathbb{Z}_p}) = H_0(\mathbb{Q}_p - \mathbb{Z}_p)$. Therefore, A is an analytic linear operator on $C(\mathbb{Z}_p, \mathbb{Q}_p)$ with compact support.

For $0 < r_1 < r_2$, define the set $D(a; r_1, r_2) = \{b \in \mathbb{K} : r_1 \leq |a - b| \leq r_2\}$.

Definition 4.28 ([54]). A function $F : D(a; r_1, r_2) \to B(\mathbb{X})$ is called an analytic operator valued function if for all $h \in \mathbb{X}^*$ and $u \in \mathbb{X}$, the function

$$z \mapsto \langle h, F(z)u \rangle = \sum_{j \in \mathbb{Z}} F_j \cdot (z - a)^j,$$

with

$$\lim_{j \to -\infty} \left| F_j \right| r_1^j = 0 \text{ and } \lim_{j \to \infty} \left| F_j \right| r_2^j = 0.$$

Lemma 4.29 ([54]). *Let $F : D(a; r_1, r_2) \to B(\mathbb{X})$ be an analytic operator valued function. Then the sequence $(S_n)_n \subset B(\mathbb{X})$ defined by*

$$S_n := \frac{1}{n} \sum_{\eta^n = 1} F(a + \Gamma \eta)$$

converges strongly as $n \to \infty$ (the limit is taken assuming that $(n, char(\mathbb{K})) = 1$ when $char(\mathbb{K}) \neq 0$) to a bounded linear operator. More precisely,

$$\lim_{n \to \infty} S_n := \int_{a, \Gamma} F(z) dz.$$

Proof. Let $h \in c_0^*(\mathbb{I}, \mathbb{K})$ and $u \in c_0(\mathbb{I}, \mathbb{K})$. Then,

$$\langle h, S_m u \rangle = \frac{1}{m} \langle h, \sum_{\eta^m = 1} F(a + \Gamma \eta) u \rangle$$

$$= \frac{1}{m} \sum_{\eta^m = 1, j \in \mathbb{Z}} F_j (\Gamma \eta)^j.$$

Thus taking the limit over m relatively prime to $char(\mathbb{K})$ if $char(\mathbb{K}) \neq 0$, one obtains

$$\lim_{m \to \infty} {}' \langle h, S_m u \rangle = \lim_{m \to \infty} {}' \frac{1}{m} \sum_{\eta^m = 1, j \in \mathbb{Z}} F_j (\Gamma \eta)^j$$

$$\langle h, (\lim_{m \to \infty} {}' S_m) u \rangle = \int_{a, \Gamma} \langle h, F(z) u \rangle dz.$$

Hence the limit of $S_m u$ exists weakly and by the uniform boundedness principle, $\lim_{m \to \infty} {}' S_m$ exists strongly as well.

Corollary 4.30 ([54]). *We have*

$$\langle h, \left(\int_{a, \Gamma} F(z) dz \right) u \rangle = \int_{a, \Gamma} \langle h, F(z) u \rangle dz.$$

4.5 Vishik Spectral Theorem

Let $\sigma \subset \mathbb{K}$ be a compact subset. An operator valued distribution is a $\mathbb{K}$-linear continuous mapping $L(\sigma) \to B(\mathbb{E}_\omega)$.

The so-called Vishik spectral theorem is given as follows:

Theorem 4.31 ([54]). *Let $A : \mathbb{E}_\omega \mapsto \mathbb{E}_\omega$ be an analytic bounded linear operator with compact spectrum. Then the following operator valued distribution $\mu_A := VR_A$ ($\mu_A(f) := f(A)$ for all $f \in B_r(\sigma)$) with support σ is given by*

$$\mu_A(f) = \sum_{i \in J} \int_{a_i, \Gamma} (z - a_i) R_A(z) f(z) dz = \sum_{i \in J} \int_{a_i, \Gamma} (z - a_i)(zI - A)^{-1} f(z) dz$$

is well-defined for all $f \in B_r(\sigma)$.

Proof. Let $0 < r_1 < r$. Let I and $\Gamma \in \mathbb{K}$ be such that for all $i \in I$,

$$\sup_{b \in \sigma_A \cap D(a_i, r)} |a_i - b| < |\Gamma| < r.$$

In addition, let $I', a'_k, \Gamma' \in \mathbb{K}$ be such that for all $k \in I'$, we have

$$\sup_{b \in \sigma_A \cap D(a'_k, r_1)} |a'_k - b| < |\Gamma'| < r_1.$$

It suffices to show that for all $f \in B_r(\sigma_A)$,

$$\sum_{i \in I} \int_{a_i, \Gamma} (z - a_i) R_A(z) f(z) dz = \sum_{k \in I'} \int_{a'_k, \Gamma} (z - a'_k) R_A(z) f(z) dz.$$

But this is true since the matrix elements on the left and right hand sides of the above equation coincide based upon Lemma 4.24 and Corollary 4.30.

4.6 Bibliographical Notes

The material of this chapter (including proofs) was taken from the following four sources: Attimu [4], Attimu and Diagana [3], Baker [7], and Vishik [54].

Chapter 5
Spectral Theory for Perturbations of Bounded Diagonal Linear Operators

Our main goal in this chapter consists of computing the spectrum of the class of bounded linear operators, $A = D + F$ where D is a diagonal operator and F is a finite rank operator. In order to achieve that, we will make extensive use of the theory of Fredholm operators and the notion of essential spectrum. A few illustrative examples will be discussed at the end of the end of this chapter. Here, we mainly follow Diagana et al. [18].

5.1 Spectral Theory for Finite Rank Perturbations of Diagonal Operators

We study the spectral analysis for classes of finite rank perturbations of diagonal Operators in the form,

$$A = D + F,$$

where D is a diagonal operator and $F = u_1 \otimes v_1 + u_2 \otimes v_2 + \ldots + u_m \otimes v_m$ is an operator of finite rank in $\mathbb{E}_\omega$.

5.1.1 Introduction

In Diarra [19], the spectral analysis of the operator $\widehat{B}$ defined on $\mathbb{E}_\omega$ and given by

$$\widehat{B}e_j = e_j + \sum_{i \neq j} \omega_i^{-1} e_i \tag{5.1}$$

was thoroughly investigated.

© The Author(s) 2016
T. Diagana, F. Ramaroson, *Non-Archimedean Operator Theory*,
SpringerBriefs in Mathematics, DOI 10.1007/978-3-319-27323-5_5

More precisely, Diarra has shown that under some suitable assumptions that the spectrum $\sigma(\widehat{B})$ of $\widehat{B}$ is given by $\sigma(\widehat{B}) = \{1\} \cup \sigma_p(\widehat{B})$, where $\sigma_p(\widehat{B})$ is the collection of eigenvalues of $\widehat{B}$. Furthermore, Diarra has shown that the eigenvalues of $\widehat{B}$ are of the form $\lambda = 1 + \alpha$, where α runs over the collection of all zeros of the function defined by

$$\varphi(\alpha) = 1 - \sum_{j \in \mathbb{N}} \frac{1}{1 + \alpha \omega_j}$$

and next made extensive use of the classical p-adic analytic functions theory to locate all the zeros of φ.

Using similar techniques as in Diarra's work, Diagana and McNeal [14, 15] computed the spectrum for elements of the class of linear operators denoted $\mathcal{D}_{per}(\mathbb{E}_\omega)$, which consists of all bounded linear operators on $\mathbb{E}_\omega$ of the form

$$\widehat{A} = D + u \otimes v, \tag{5.2}$$

where the diagonal operator D is defined by $De_j = \lambda_j e_j$ with $\lambda = (\lambda_j)_{j \in \mathbb{N}} \subset \mathbb{K}$ being a sequence and $u \otimes v$ for each $u = (\alpha_j)_{j \in \mathbb{N}}, v = (\beta_j)_{j \in \mathbb{N}} \in \mathbb{E}_\omega$ with $\alpha_j, \beta_j \in \mathbb{K} - \{0\}$ for all $j \in \mathbb{N}$, is the rank-one linear operator defined by $(u \otimes v)(w) = \langle v, w \rangle u$ for each $w \in \mathbb{E}_\omega$. Namely, they have shown that if $\widehat{A} = D + u \otimes v$ belongs to $\mathcal{D}_{per}(\mathbb{E}_\omega)$, then its spectrum, under some suitable assumptions, is given by

$$\sigma(\widehat{A}) = \{\theta_j\}_{j \geq 0} \cup \sigma_p(\widehat{A}),$$

where $\sigma_p(\widehat{A})$ is the set of eigenvalues of $\widehat{A}$ and $\theta = (\theta_j)_{j \in \mathbb{N}}$ with $\theta_j = \lambda_j + \omega_j \alpha_j \beta_j$ for all $j \in \mathbb{N}$.

It should be mentioned that the operator $\widehat{B}$ given in Eq. (5.1) is a particular element of $\mathcal{D}_{per}(\mathbb{E}_\omega)$. Indeed, assuming that $\omega_j^{-1} \to 0$ as $j \to \infty$, taking $\widehat{u} = \widehat{v} = (\omega_j^{-1})_{j \in \mathbb{N}} \in \mathbb{E}_\omega$ and letting $\widehat{D}e_j = (1 - \omega_j^{-1})e_j$ for $j \in \mathbb{N}$ in Eq. (5.2), one can easily see that $\widehat{B} = \widehat{D} + \widehat{u} \otimes \widehat{v}$ with $\lambda = (1 - \omega_j^{-1})_{j \in \mathbb{N}}$.

In this chapter, we will make extensive use of Fredholm operator theory in the non-archimedean setting as well as the essential spectrum to study the spectral analysis of operators of the form

$$T = D + K,$$

where D is a (bounded) diagonal linear operator and K is a completely continuous linear operator on $\mathbb{E}_\omega$. Next, we deduce the spectral analysis for the class of finite rank perturbations of diagonal operators given by

$$T = D + F, \tag{5.3}$$

where D is a diagonal operator and $F = u_1 \otimes v_1 + u_2 \otimes v_2 + \ldots + u_m \otimes v_m$ is an operator of rank at most m with both $u_k = (\alpha_j^k)_{j \in \mathbb{N}}$ and $v_k = (\beta_j^k)_{j \in \mathbb{N}}$ belonging to $\mathbb{E}_\omega \setminus \{0\}$ for $k = 1, 2, \ldots, m$.

In order to illustrate the abstract results of this chapter, a few examples will be discussed (see Examples 5.22 and 5.24).

5.1.2   Spectral Analysis for the Class of Operators $T = D + K$

In this subsection we study the spectral analysis for perturbations of completely continuous linear operators by diagonal operators. Namely, we study the spectral theory of the class of operators of the form

$$T = D + K,$$

where $D : \mathbb{E}_\omega \mapsto \mathbb{E}_\omega$ is a diagonal operator defined by $De_j = \lambda_j e_j$ for all $j \in \mathbb{N}$ where $\lambda = (\lambda_j)_{j \in \mathbb{N}} \subset \mathbb{K}$ is a bounded sequence and $K : \mathbb{E}_\omega \mapsto \mathbb{E}_\omega$ is a completely continuous linear operator.

We will need the following set in the sequel:

$$\Phi_0(\mathbb{E}_\omega) := \left\{ A \in \Phi(\mathbb{E}_\omega) : \chi(A) = 0 \right\}.$$

Theorem 5.1. *If $A \in \Phi(\mathbb{E}_\omega)$ and $K \in C(\mathbb{E}_\omega)$, then $A + K \in \Phi(\mathbb{E}_\omega)$ with*

$$\chi(A + K) = \chi(A).$$

Proof. We use the notations of Sect. 3.5. So let $\mathbb{E}_\omega = \mathbb{E}_{\omega 0} \oplus N(A)$ and $\mathbb{E}_\omega = \mathbb{E}_{\omega 0} \oplus R(A)$. Define the linear operators $\tilde{A}$ and $\tilde{K}$ on $\mathbb{E}_{\omega 0} \times \mathbb{E}_{\omega 0}$ by

$$\tilde{A}(x_0, y_0) = Ax_0 + y_0 \text{ and } \tilde{K} = Kx_0 + y_0.$$

Using the facts that K is completely continuous and that $\mathbb{E}_{\omega 0}$ is a finite dimensional subspace it follows that $\tilde{K}$ is completely continuous. Now since $(\tilde{A} + \tilde{K})(x_0, 0) = (A + K)x_0$, using Lemma 3.41 it follows that $A + K$ is Fredholm if and only if $\tilde{A} + \tilde{K}$ is. Since $\tilde{A}$ is bijective, one can write $\tilde{A} + \tilde{K} = \tilde{A}(I + \tilde{A}^{-1}\tilde{K})$. Obviously $\tilde{A}^{-1}\tilde{K}$ is completely continuous. From Proposition 3.46 it follows that $I + \tilde{A}^{-1}\tilde{K}$ is Fredholm. Consequently, $A + K \in \Phi(\mathbb{E}_\omega)$.

Theorem 5.2. *If $A \in B(\mathbb{E}_\omega)$, then for all $K \in C(\mathbb{E}_\omega)$, we have $\sigma_e(A + K) = \sigma_e(A)$.*

Proof. The proof is an immediate consequence of Theorem 5.1. Indeed, if λ does not belong to $\sigma_e(A)$, then $\lambda I - A$ belongs to $\Phi(\mathbb{E}_\omega)$ with $\chi(\lambda I - A) = 0$. Therefore, $\lambda I - A - K$ belongs to $\Phi(\mathbb{E}_\omega)$ with $\chi(\lambda I - A - K) = 0$ for all $K \in C(\mathbb{E}_\omega)$.

Corollary 5.3. *For every $K \in C(\mathbb{E}_\omega)$, we have $\sigma_e(D + K) = \sigma_e(D)$.*

Proposition 5.4. *If $T = D + K$ where $K \in C(\mathbb{E}_\omega)$, then its spectrum $\sigma(T)$ is given by*

$$\sigma(T) = \sigma_e(D) \cup \sigma_p(T).$$

Proof. This is an immediate consequence of Corollary 5.3 and the fact that $\sigma(T) = \sigma_p(T) \cup \sigma_e(T)$.

Corollary 5.5. *If $T = D + u_1 \otimes v_1 + u_2 \otimes v_2 + \ldots + u_m \otimes v_m$, then its spectrum $\sigma(T)$ is given by $\sigma(T) = \sigma_e(D) \cup \sigma_p(T)$.*

Lemma 5.6. *If $A \in \Phi_0(\mathbb{E}_\omega)$ and $K \in C(\mathbb{E}_\omega)$, then the linear operator $A + K$ is invertible if and only if $N(A + K) = \{0\}$.*

Proof. Since $A \in \Phi(\mathbb{E}_\omega)$ with index $\chi(A) = 0$ it follows, by using Theorem 5.1, that $A + K$ belongs to $\Phi(\mathbb{E}_\omega)$ with index $\chi(A + K) = \chi(A) = 0$. In other words,

$$\eta(A + K) = \delta(A + K).$$

Now if $A + K$ is invertible, then $N(A + K) = \{0\}$. Conversely, if $N(A + K) = \{0\}$, then $0 = \eta(A + K) = \delta(A + K)$ and hence $A + K$ must be surjective, that is, $A + K$ is invertible.

In the rest of this chapter, we suppose that there exists $d_k \neq 0$ for $k = 1, 2, \ldots, m$ such that

$$\langle u_k, u_l \rangle = d_k \delta_{kl} \tag{5.4}$$

for $k, l = 1, 2, \ldots, m$.

Lemma 5.7. *Consider the finite rank operator*

$$F = \sum_{k=1}^{m} u_k \otimes v_k,$$

where $u_k = (\alpha_j^k)_{j \in \mathbb{N}}$, $v_k = (\beta_j^k)_{j \in \mathbb{N}} \in \mathbb{E}_\omega$ with $\alpha_j^k, \beta_j^k \in \mathbb{K} \setminus \{0\}$ for each $k = 1, 2, \ldots, m$ and $j \in \mathbb{N}$. Then, the operator $I - F$ (respectively, $I + F$) is invertible if and only if $\det P \neq 0$ (respectively, $\det Q \neq 0$), where P (respectively, Q) is the $m \times m$ square matrix given by $P = (a_{ij})_{i,j=1,\ldots,m}$ (respectively, given by $Q = (b_{ij})_{i,j=1,\ldots,m}$) with $a_{ij} = \delta_{ij} - \langle u_j, v_i \rangle$ (respectively, $b_{ij} = \delta_{ij} + \langle u_j, v_i \rangle$).

Proof. Using Lemma 5.6 it follows that the operator $I - F$ (respectively, $I + F$) is invertible if and only if $N(I - F) = \{0\}$ (respectively, $N(I + F) = \{0\}$). To complete the proof it is sufficient to show that $N(I - F) = \{0\}$ (respectively, $N(I + F) = \{0\}$) if only if $\det P \neq 0$ (respectively, $\det Q \neq 0$). For that, let $w \in \mathbb{E}_\omega$ such that $(I - F)w = 0$. Equivalently,

$$w - \sum_{k=1}^{m} \langle v_k, w \rangle u_k = 0. \tag{5.5}$$

Now taking the inner product of Eq. (5.5) with respectively $v_1, v_2, \ldots,$ and v_m, we obtain the following system of equations

$$P \begin{pmatrix} \langle w, v_1 \rangle \\ \langle w, v_2 \rangle \\ . \\ . \\ \langle w, v_m \rangle \end{pmatrix} = \begin{pmatrix} 0 \\ 0 \\ . \\ . \\ 0 \end{pmatrix}. \tag{5.6}$$

If we suppose that $N(I - F) \neq \{0\}$, then

$$w = \sum_{k=1}^{m} \langle v_k, w \rangle u_k \neq 0$$

and hence at least one of the following scalars $\langle w, v_1 \rangle, \langle w, v_2 \rangle, \ldots, \langle w, v_m \rangle$ is nonzero. Consequently, Eq. (5.6) has at least one nontrivial solution which yields $\det P = 0$.

Conversely, if $\det P = 0$, there exist some scalars $\xi_1, \ldots, \xi_m$ not all zeros, such that with $\xi = (\xi_1, \ldots, \xi_m)^t$ we have

$$P \begin{pmatrix} \xi_1 \\ \xi_2 \\ . \\ . \\ \xi_m \end{pmatrix} = \begin{pmatrix} 0 \\ 0 \\ . \\ . \\ 0 \end{pmatrix}.$$

Guided by Eq. (5.5), we take $w = \sum_{k=1}^{m} \xi_k u_k$ and obtain $(I - F)w = 0$. Now $w \neq 0$. If not, then, $0 = \langle w, u_j \rangle = d_j \xi_j$ for $j = 1, 2, \ldots, m$ which yields $\xi_j = 0$ for $j = 1, 2, \ldots, m$, and that contradicts the fact that some of the ξ_j are nonzero. In view of the above, $N(I - F) \neq 0$. The proof for $I + F$ is similar to that of $I - F$ and hence is omitted.

Proposition 5.8. *Consider the finite rank operator*

$$F = \sum_{k=1}^{m} u_k \otimes v_k,$$

where $u_k = (\alpha_j^k)_{j \in \mathbb{N}}$, $v_k = (\beta_j^k)_{j \in \mathbb{N}} \in \mathbb{E}_\omega$ with $\alpha_j^k, \beta_j^k \in \mathbb{K} \setminus \{0\}$ for each $k = 1, 2, \ldots, m$ and $j \in \mathbb{N}$. Then, the spectrum of F is given by

$$\sigma(F) = \left\{ \lambda \in \mathbb{K} \setminus \{0\} : \det P(\lambda) = 0 \right\} \cup \left\{ 0 \right\},$$

where $P(\lambda)$ is the $m \times m$ square matrix given by $P(\lambda) = (a_{ij}(\lambda))_{i,j=1,\ldots,m}$ with

$$a_{ij}(\lambda) = \lambda \delta_{ij} - \langle u_j, v_i \rangle.$$

Proof. Consider the operator $\lambda I - F$. Clearly, $\lambda = 0$ is necessarily in the spectrum F as F is not invertible. Now, suppose $\lambda \neq 0$. Then $\lambda I - F = \lambda(I - F_\lambda)$ where $F_\lambda = \lambda^{-1} F$ is a finite rank operator. It is then clear that $\lambda I - F$ is invertible if and only if $I - F_\lambda$ is in which case $(\lambda I - F)^{-1} = \lambda^{-1}(I - F_\lambda)^{-1}$. Now using Lemma 5.6 it follows that $I - F_\lambda$ is invertible if and only if $N(I - F_\lambda) = \{0\}$. Following along the same lines as in the proof of Lemma 5.7 it follows that $N(I - F_\lambda) = \{0\}$ if and only if $\lambda \in \mathbb{K} \setminus \{0\}$ and $\det P(\lambda) \neq 0$. This completes the proof. $\quad\blacksquare$

5.1.3   Spectral Analysis for the Class of Operators $T = D + F$

In this subsection, we make extensive use of the results of Sect. 5.1.2 to study the spectral analysis of operators of the operators of the form

$$T = D + F,$$

where $D : \mathbb{E}_\omega \mapsto \mathbb{E}_\omega$ is a diagonal operator defined by $De_j = \lambda_j e_j$ with $\lambda = (\lambda_j)_{j \in \mathbb{N}} \subset \mathbb{K}$ being a bounded sequence and F is an operator of finite rank defined by

$$F = \sum_{k=1}^m u_k \otimes v_k,$$

with $u_k = (\alpha_j^k)_{j \in \mathbb{N}}$, $v_k = (\beta_j^k)_{j \in \mathbb{N}} \in \mathbb{E}_\omega$ with $\alpha_j^k, \beta_j^k \in \mathbb{K} \setminus \{0\}$ for each $k = 1, 2, \ldots, m$ and $j \in \mathbb{N}$.

Let $\mathcal{D}_{per}^0(\mathbb{E}_\omega)$ denote the collection of all linear operators of the form

$$T = D + F = D + u_1 \otimes v_1 + u_2 \otimes v_2 + \ldots + u_m \otimes v_m$$

such $u_k = (\alpha_j^k)_{j \in \mathbb{N}}$, $v_k = (\beta_j^k)_{j \in \mathbb{N}} \in \mathbb{E}_\omega$ with $\alpha_j^k, \beta_j^k \in \mathbb{K} \setminus \{0\}$ for each $k = 1, 2, \ldots, m$ and $j \in \mathbb{N}$.

Lemma 5.9. *If $T = D + F$ belongs to $\mathcal{D}_{per}^0(\mathbb{E}_\omega)$ where D is a bounded diagonal operator on $\mathbb{E}_\omega$ and $F = \sum_{k=1}^m u_k \otimes v_k$. Then $\lambda \in \sigma_p(T)$ if and only if*

(a) $\lambda \notin \sigma_p(D) = \left\{ \lambda_j : j \in \mathbb{N} \right\}$; i.e., $\lambda \neq \lambda_j$ for all $j \in \mathbb{N}$; and

(b) $\det M(\lambda) = 0$ where $M(\lambda) = (c_{ij}(\lambda))$ is an $m \times m$ square matrix with entries:

$$c_{ij}(\lambda) = \delta_{ij} + \langle C_\lambda u_j, v_i \rangle$$

for $i, j = 1, 2, \ldots, m$ with $C_\lambda := (D - \lambda I)^{-1}$.

Proof. Suppose $\lambda \in \sigma_p(T)$. Thus there exists $0 \neq w \in \mathbb{E}_\omega$ such that $Tw = \lambda w$. Equivalently,

$$(\lambda I - D)w = Fw = \sum_{k=1}^{m} \langle v_k, w \rangle u_k. \tag{5.7}$$

Clearly, all the expressions $\langle v_k, w \rangle$ are nonzero for $k = 1, 2, \ldots, m$. If not, we will get $(\lambda I - D)w = 0$ with $w \neq 0$. That is, $\lambda \in \sigma_p(D)$ and hence there exists $j_0 \in \mathbb{N}$ such that $\lambda = \lambda_{j_0}$, $w = a e_{j_0}$ with $a \in \mathbb{K} \setminus \{0\}$ and $\omega_{j_0} \beta_{j_0}^k = \langle v_k, e_{j_0} \rangle = a^{-1} \langle v_k, w \rangle = 0$ yields $\beta_{j_0}^k = 0$ for $k = 1, 2, \ldots, m$, which contradicts the fact that $\beta_j^k \in \mathbb{K} \setminus \{0\}$ for each $k = 1, 2, \ldots, m$ and $j \in \mathbb{N}$. Consequently, $Fw = (\lambda I - D)w \neq 0$ and hence $u_k \in R(\lambda I - D)$ for $k = 1, 2, \ldots, m$ and $\lambda \notin \sigma_p(D)$.

Clearly, Eq. (5.7) is equivalent to

$$w + \sum_{k=1}^{m} \langle v_k, w \rangle C_\lambda u_k = 0. \tag{5.8}$$

Taking the inner product of Eq. (5.8) with respectively $v_1, v_2, \ldots,$ and v_m, then we obtain the following system of equations

$$M(\lambda) \begin{pmatrix} \langle w, v_1 \rangle \\ \langle w, v_2 \rangle \\ . \\ . \\ \langle w, v_m \rangle \end{pmatrix} = \begin{pmatrix} 0 \\ 0 \\ . \\ . \\ 0 \end{pmatrix}. \tag{5.9}$$

Using the fact that at least one of the following numbers $\langle w, v_1 \rangle$, $\langle w, v_2 \rangle$, $\ldots,$ $\langle w, v_m \rangle$ is nonzero it follows that Eq. (7.4) has at least one nontrivial solution which yields $\det M(\lambda) = 0$.

Suppose (a)–(b) are true, that is, $\lambda \notin \sigma_p(D)$ and $\det M(\lambda) = 0$. Then there exist some scalars $\xi_1, \ldots, \xi_m$ not all zeros, such that with $\xi = (\xi_1, \ldots, \xi_m)^t$ we have

$$M(\lambda)\begin{pmatrix}\xi_1\\\xi_2\\.\\.\\.\\\xi_m\end{pmatrix}=\begin{pmatrix}0\\0\\.\\.\\.\\0\end{pmatrix}. \tag{5.10}$$

Guided by Eq. (7.3), we take $w = -\sum_{k=1}^{m}\xi_k C_\lambda u_k$ and obtain $(T - \lambda I)w = 0$. Now $w \neq 0$. For that, let us show that $C_\lambda^{-1}w \neq 0$ which, by using C_λ, yields $w \neq 0$. If $C_\lambda^{-1}w = 0$ it follows that $0 = \langle C_\lambda^{-1}w, u_j\rangle = -\xi_j d_j$ for $j = 1, 2, \ldots, m$ which yields $\xi_j = 0$ for $j = 1, 2, \ldots, m$ and that contradicts the fact that some of the ξ_j are nonzero. In view of the above, $N(T - \lambda I) \neq 0$. That is, $\lambda \in \sigma_p(T)$.

Corollary 5.10. *Let $T = D + F$ belong to $\mathcal{D}_{per}^0(\mathbb{E}_\omega)$ and let $\lambda \in \rho(D)$. Then, $\lambda \in \sigma_p(T)$ if and only if $\det M(\lambda) = 0$ where $M(\lambda) = (c_{ij}(\lambda))$ with $c_{ij}(\lambda) = \delta_{ij} + \langle C_\lambda u_j, v_i\rangle$ for $i, j = 1, 2, \ldots, m$ with $C_\lambda := (D - \lambda I)^{-1}$.*

Proof. Since $\lambda \in \rho(D)$ it follows that $\lambda \notin \sigma_p(D)$. Using the similar ideas as in the proof of Lemma 5.9 it follows easily that $\det M(\lambda) = 0$. The converse is also clear.

Corollary 5.11. *Suppose $T = D + F$ belongs to $\mathcal{D}_{per}^0(\mathbb{E}_\omega)$ where $F = \sum_{k=1}^{m} u_k \otimes v_k$, then the eigenvalues of T are given by*

$$\sigma_p(T) = \Big\{\lambda \in \rho(D) : \det M(\lambda) = 0\Big\}.$$

Proof. Corollary 5.10 yields $\Big\{\lambda \in \rho(D) : \det M(\lambda) = 0\Big\} \subset \sigma_p(T)$. For the other inclusion, let us then assume that $\lambda \notin \sigma_e(D) = \sigma_e(T)$. To complete the proof, we have to show that $\det M(\lambda) = 0$. For that, suppose $\lambda \in \sigma_p(T)$. If $\lambda \in \sigma_p(T) \cap \overline{\sigma_p(D)}$, then using Lemma 5.9 it follows that $\lambda \notin \sigma_p(D)$. Consequently, one must have $\lambda \in \sigma_e(D) = \sigma_e(T)$, which is impossible by assumption. Thus, one must have $\lambda \in \sigma_p(T) \setminus \overline{\sigma_p(D)}$, which by Corollary 5.10 yields $\det M(\lambda) = 0$.

We have:

Corollary 5.12. *Suppose $T = D + F$ belongs to $\mathcal{D}_{per}^0(\mathbb{E}_\omega)$ where $F = \sum_{k=1}^{m} u_k \otimes v_k$, then the spectrum $\sigma(T)$ of T is given by*

$$\sigma(T) = \Big\{\lambda \in \rho(D) : \det M(\lambda) = 0\Big\} \cup \sigma_e(D).$$

Using the fact that $T^* = D + F^*$ where $F^* : \mathbb{E}_\omega \mapsto \mathbb{E}_\omega$ is the finite rank operator given by $F^* = \sum_{k=1}^{m} v_k \otimes u_k$ and Corollary 5.12, we obtain:

Corollary 5.13. *If $T = D + F$ belongs to $\mathcal{D}_{per}^0(\mathbb{E}_\omega)$ where $F = \sum_{k=1}^{m} u_k \otimes v_k$ such that there exists $s_k \neq 0$ for $k = 1, 2, \ldots, m$ and*

$$\langle v_k, v_l\rangle = s_k \delta_{kl}$$

for $k, l = 1, 2, \ldots, m$, then the spectrum $\sigma(T^*)$ of the adjoint T^* of T is given by

$$\sigma(T^*) = \left\{ \lambda \in \rho(D) : \det N(\lambda) = 0 \right\} \cup \sigma_e(D),$$

where $N(\lambda)$ being the $m \times m$ square matrix $N(\lambda) = (d_{ij}(\lambda))_{i,j=1,2,\ldots,m}$ whose coefficients are given by

$$d_{ij}(\lambda) := \delta_{ij} + \langle (D - \lambda I)^{-1} v_j, u_i \rangle$$

for $i, j = 1, 2, \ldots, m$.

One should indicate that Eq. (5.4) is not needed in Corollary 5.13.

5.2 Computation of $\sigma_e(D)$

Our main objective in this subsection consists of computing $\sigma_e(D)$, the essential spectrum of the diagonal operator D. For that we need a few notations. Let $\Lambda = \{\lambda_j : j \in \mathbb{N}\}$. For each $\lambda \in \Lambda$, let $I_\lambda = \{k \in \mathbb{N} : \lambda_k = \lambda\}$ and let $r_\lambda = |I_\lambda|$. Let $\Lambda^* = \{\lambda \in \Lambda : r_\lambda < \infty\}$ and $\Lambda' = \{\lambda \in \Lambda : \lambda \text{ is an accumulation point of } \Lambda\}$.

To each $\lambda \in \Lambda$, we associate the subspace V_λ of $\mathbb{E}_\omega$ defined by

$$V_\lambda = \left\{ \sum_i v_i e_i : v_i = 0 \text{ for all } i \in I_\lambda \right\}.$$

Then V_λ is a closed subspace of $\mathbb{E}_\omega$.

Proposition 5.14. *For each $\lambda \in \Lambda$, let W_λ be the subspace algebraically generated by $\{e_k : k \notin I_\lambda\}$ and $\overline{W_\lambda}$ its closure in $\mathbb{E}_\omega$. Then,*

$$\overline{W_\lambda} = V_\lambda, \text{ and } W_\lambda \subset R(\lambda I - D) \subset V_\lambda.$$

Proof. Clearly, $W_\lambda \subset V_\lambda$ which implies the first assertion. For the second one, we observe that for $k \notin I_\lambda$, $e_k = (\lambda I - D)\left(\frac{e_k}{\lambda - \lambda_k}\right)$ which yields the first inclusion. The second inclusion is clear.

Theorem 5.15. $\sigma_e(D) = \left(\overline{\Lambda} \setminus \Lambda^*\right) \cup (\Lambda^* \cap \Lambda')$, *where $\overline{\Lambda}$ is the closure of Λ in $\mathbb{K}$.*

We will prove Theorem 5.15 through a series of technical lemmas.

Lemma 5.16. $\overline{\Lambda} \setminus \Lambda^* \subset \sigma_e(D) \subset \left(\overline{\Lambda} \setminus \Lambda^*\right) \cup (\Lambda^* \cap \Lambda')$.

Proof. For the first inclusion, let $\lambda \in \overline{\Lambda} \setminus \Lambda^*$. Now, $\overline{\Lambda} = \sigma(D) = \sigma_p(D) \cup \sigma_e(D) = \Lambda \cup \sigma_e(D)$. If $\lambda \in \sigma_e(D)$, then we are done. Otherwise, $\lambda \in \Lambda$, then r_λ is infinite and hence $\dim(N(\lambda I - D)) = \infty$. Therefore, $\lambda I - D$ is not a Fredholm operator of index zero, hence $\lambda \in \sigma_e(D)$.

For the next inclusion, let $\lambda \in \sigma_e(D)$. Suppose $\lambda \notin \Lambda$, then $\lambda \in \overline{\Lambda} \setminus \Lambda^*$. Suppose $\lambda \in \overline{\Lambda}$ and $r_\lambda = \infty$, then $\lambda \in \overline{\Lambda} \setminus \Lambda^*$. Suppose $\lambda \in \Lambda$ and $r_\lambda < \infty$. Assume that $\lambda \notin \Lambda'$.

(a) *Claim*: $R(\lambda I - D)$ is closed. We show that $R(\lambda I - D) = V_\lambda$. For this, let $v = \sum_i v_i e_i \in V_\lambda$ then $v_i = 0$ for all $i \in I_\lambda$ where I_λ is finite. Then,

$$v = \sum_{k \notin I_\lambda} v_k e_k \text{ and } \lim_k \left\| v_k \right\| \left\| e_k \right\| = 0.$$

Now consider $w = \sum_{k \notin I_\lambda} \dfrac{v_k e_k}{\lambda - \lambda_k}$. Then, $\inf |\lambda - \lambda_k| > 0$ as λ is not an accumulation point of Λ, and therefore

$$\lim_k \left| \frac{v_k}{\lambda - \lambda_k} \right| \left\| e_k \right\| = 0$$

and hence $w \in \mathbb{E}_\omega$ and thus $v = (\lambda I - D)(w)$. Using Proposition 5.14, it follows that $R(\lambda I - D) = V_\lambda$.

(b) *Claim*: $\dim(\mathbb{E}_\omega / R(\lambda I - D)) = r_\lambda < \infty$. Indeed, $\mathbb{E}_\omega / R(\lambda I - D) = \mathbb{E}_\omega / V_\lambda$ but $\dim(\mathbb{E}_\omega / V_\lambda) = r_\lambda$.

(c) *Claim*: $\chi(\lambda I - D) = 0$. Indeed, this index is equal to

$$\dim(N(\lambda I - D)) - \dim(\mathbb{E}_\omega / R(\lambda I - D)) = r_\lambda - r_\lambda = 0.$$

Now we see that $r_\lambda < \infty$ and claims (a)–(c) imply that $\lambda \in \Phi_0(\mathbb{E}_\omega)$ which is against our hypothesis that $\lambda \in \sigma_e(D)$. Therefore, if r_λ is finite, then $\lambda \in \Lambda'$.

Putting everything together, we find that

$$\sigma_e(D) \subset \left(\overline{\Lambda} \setminus \Lambda^* \right) \cup \left(\Lambda^* \cap \Lambda' \right)$$

and the proof is complete.

To complete the proof of Theorem 5.15, we need to prove the following lemma.

Lemma 5.17. $\left(\Lambda^* \cap \Lambda' \right) \subset \sigma_e(D)$.

Proof. We prove this by contradiction. Let $\lambda \in \Lambda^* \cap \Lambda'$ and we suppose that $\lambda \notin \sigma_e(D)$. This assumption implies that $R(\lambda I - D)$ is closed and hence

$$R(\lambda I - D) = V_\lambda.$$

The contradiction we seek is this: we construct a vector $v \in V_\lambda$ which cannot be in $R(\lambda I - D)$.

We now continue the proof of Lemma 5.17. The proof of the following lemma is clear.

Lemma 5.18. *For $\rho \in \mathbb{K}$ such that $|\rho| > 1$, then for any $x \in \mathbb{R}^+$, there exists $n \in \mathbb{Z}$ such that*

$$\left|\rho\right|^n \leq x < \left|\rho\right|^{n+1}.$$

Corollary 5.19. *For every k, there exists $n_k \in \mathbb{Z}$ such that*

$$1 \leq \frac{\left|\omega_k\right|^{1/2}}{\left|\rho\right|^{n_k}} < \left|\rho\right|.$$

Since $\lambda \in \Lambda'$, there exists a subsequence $\{\lambda_{k_j}\} \subset \Lambda$ such that

$$\left|\lambda - \lambda_{k_j}\right| \to 0 \text{ as } j \to \infty, \text{ and } \lambda_{k_j} \neq \lambda \text{ for all } j.$$

Lemma 5.20. *Let $v = \sum\limits_i v_i e_i$ be defined by*

$$v_i = \begin{cases} 0, & \text{if } i \notin \{k_j : j \in \mathbb{N}\} \\ \left(\lambda - \lambda_{k_j}\right) \rho^{-n_{k_j}}, & \text{if } i = k_j \text{ for some } j \end{cases}$$

Then $v \in V_\lambda$ but $v \notin R(\lambda I - D)$.

Proof. It is clear that $v \in E_\omega$ because of Corollary 5.19 and the fact that $\left|\lambda - \lambda_{k_j}\right| \to 0$ as $j \to \infty$. Moreover, since we know that $I_\lambda \cap I_{k_j} = \varnothing$ for all j, we see that $v \in V_\lambda$.

Now suppose that $v \in R(\lambda I - D)$ then there exists $u = \sum\limits_i u_i e_i \in E_\omega$ such that $(\lambda I - D)(u) = v$. This implies that

$$\sum_i (\lambda - \lambda_i) u_i e_i = \sum_i v_i e_i.$$

Hence, for every j,

$$\left(\lambda - \lambda_{k_j}\right) u_{k_j} = v_{k_j} = \left(\lambda - \lambda_{k_j}\right) \rho^{-n_{k_j}}$$

and consequently,

$$u_{k_j} = \frac{1}{\rho^{n_{k_j}}} \text{ for all } j.$$

However,

$$\lim_j \left| u_{k_j} \right| \left\| e_{k_j} \right\| = \lim_j \frac{\left| \omega_{k_j} \right|^{1/2}}{\rho^{n_{k_j}}}$$

and by Corollary 5.19, this limit is $\neq 0$ which contradicts the fact that $u = \sum_i u_i e_i \in \mathbb{E}_\omega$. This ends the proof of Lemma 5.17 and therefore, that of Theorem 5.15 also.

5.3   Spectrum of $T = D + F$

Using Theorem 5.15 we obtain the main result of this chapter as follows:

Theorem 5.21. *If* $T = D + F$ *belongs to* $\mathcal{D}_{per}^0(\mathbb{E}_\omega)$ *where* $De_k = \lambda_k e_k$ *for all* $k \in \mathbb{N}$ *and* $F = \sum_{k=1}^m u_k \otimes v_k$, *then the spectrum* $\sigma(T)$ *of* T *is given by*

$$\sigma(T) = \left\{ \lambda \in \rho(D) : \det M(\lambda) = 0 \right\} \cup \left[\left(\overline{\Lambda} \setminus \Lambda^* \right) \cup \left(\Lambda^* \cap \Lambda' \right) \right]$$

where $M(\cdot)$ *is the* $m \times m$ *matrix given in Lemma 5.9.*

5.4   Examples

In this subsection $M(\cdot)$ stands for the $m \times m$ matrix given in Lemma 5.9-(c).

Example 5.22 (Rank-One Perturbation of a Diagonal Operator). Consider the example studied by Diarra [19]. Indeed, let $\omega = (\omega_j)_{j \in \mathbb{N}} \subset \mathbb{K} \setminus \{0\}$ be a non constant sequence. Moreover, suppose $\omega_0 = 1$ and $\omega_j^{-1} \to 0$ as $j \to \infty$. Moreover, $|\omega_j| > 1$ for all $j = 1, 2, \ldots$ and $|\omega_j| < |\omega_{j+1}|$ for all $j \in \mathbb{N}$.

Let $u = v = (\omega_j^{-1})_{j \in \mathbb{N}} \in \mathbb{E}_\omega$ and define D as follows:

$$De_j = (1 - \omega_j^{-1}) e_j$$

for $j \in \mathbb{N}$.

Since $\omega = (\omega_j)_{j \in \mathbb{N}}$ is a non constant sequence it follows that each eigenvalue $\lambda_j = 1 - \omega_j^{-1}$ of D is of multiplicity 1. Moreover, $\lambda_j \to 1$ as $j \to \infty$.

Let us show that assumption (5.4) holds. This amounts to showing that u is not isotropic, that is,

$$\langle u, u \rangle = 1 + \sum_{j=1}^{\infty} \omega_j \omega_j^{-1} \omega_j^{-1} = 1 + \sum_{j=1}^{\infty} \omega_j^{-1} \neq 0.$$

Indeed, suppose $1 + \sum_{j=1}^{\infty} \omega_j^{-1} = 0$ which yields $\sum_{j=1}^{\infty} \omega_j^{-1} = -1$, that is,

$$\left| \sum_{j=1}^{\infty} \omega_j^{-1} \right| = 1.$$

Now

$$\left| \sum_{j=1}^{N} \omega_j^{-1} \right| \leq \max_{j=1,2,\dots,N} \left| \omega_j^{-1} \right|$$

$$< \left| \omega_1^{-1} \right|$$

which yields

$$1 = \left| \sum_{j=1}^{\infty} \omega_j^{-1} \right| \leq \left| \omega_1^{-1} \right| < 1$$

and this is a contradiction and hence $\langle u, u \rangle \neq 0$.

The point and the essential spectrums of the diagonal operator D are given as follows:

$$\sigma_p(D) = \Lambda = \left\{ 1 - \omega_j^{-1} : j \in \mathbb{N} \right\} \text{ and } \sigma_e(D) = \left\{ 1 \right\}.$$

Clearly, $\det M(\lambda) = 0$ for $\lambda \in \mathbb{K} \setminus \overline{\sigma_p(D)}$ is equivalent to the following equation,

$$1 - \sum_{j=0}^{\infty} \frac{1}{\omega_j(\lambda - 1) + 1} = 0.$$

As a consequence of Theorem 5.21 we have.

Theorem 5.23. *Consider the bounded linear operator T on $\mathbb{E}_\omega$ defined by*

$$T e_j = e_j + \sum_{i \neq j} \omega_i^{-1} e_i, \text{ for all } j \in \mathbb{N}.$$

Then the spectrum of T is given by

$$\sigma(T) = \left\{ \lambda \in \mathbb{K} \setminus \left(\{1 - \omega_j^{-1}\}_{j \in \mathbb{N}} \cup \{1\} \right) : \varphi(\lambda) = 0 \right\} \cup \{1\},$$

where the function $\varphi : \mathbb{K} \setminus \left(\{1 - \omega_j^{-1}\}_{j \in \mathbb{N}} \cup \{1\} \right) \mapsto \mathbb{K}$ *is defined by*

$$\varphi(\lambda) := 1 - \sum_{j=0}^{\infty} \frac{1}{\omega_j(\lambda - 1) + 1}.$$

One should point out that Theorem 5.23 was obtained by Diarra [19]. Further, Diarra computed all the zeros of the function φ.

Example 5.24 (Rank m Perturbation of a Diagonal Operator). Let $p \geq 2$ be a prime and let $\mathbb{K} = \mathbb{Q}_p$ be equipped with the usual p-adic absolute value $|\cdot|_p$ and let $\mathbb{E}_\omega$ be the corresponding p-adic Hilbert space over $\mathbb{Q}_p$. Let $\lambda : \mathbb{N} \to \mathbb{Z}$ be a bijection. Since $\mathbb{Z}$ is countable, such a bijection exists. First note that the closure of $\mathbb{Z}$ in $\mathbb{Q}_p$ is $\mathbb{Z}_p$ the ring of p-adic integers.

Let $\varepsilon \in \{-1, 1\}$. Recall

$$\mathbb{Z}_p = \left\{ \sum_{k=0}^{\infty} a_k p^k : \ 0 \leq a_k \leq p - 1 \right\}$$

and

$$\mathbb{Z} = \left\{ \sum_{k=0}^{N} \varepsilon a_k p^k : \ 0 \leq a_k \leq p - 1, \ N \in \mathbb{N} \right\}.$$

Consider the diagonal operator $D : \mathbb{E}_\omega \to \mathbb{E}_\omega$ defined by

$$D\left((u_i)_{i \in \mathbb{N}} \right) := \sum_{i=0}^{\infty} \lambda(i) u_i e_i.$$

First of all, note that the linear operator D defined above is bounded as

$$\|D\| = \sup_{i \in \mathbb{N}} \frac{\|De_i\|}{\|e_i\|} = \sup_{i \in \mathbb{N}} \frac{|\lambda(i)|_p \|e_i\|}{\|e_i\|} = \sup_{i \in \mathbb{N}} |\lambda(i)|_p \leq 1.$$

Moreover, each eigenvalue of D is of multiplicity 1.

Now, $\Lambda = \left\{\lambda\,(i) : i \in \mathbb{N}\right\} = \mathbb{Z}$. Consequently,

(a) $\sigma(D) = \overline{\Lambda} = \mathbb{Z}_p$.
(b) $\sigma_p(D) = \Lambda = \mathbb{Z}$.
(c) $\sigma_e(D) = \mathbb{Z}_p$.

As a consequence of Theorem 5.21 we have.

Theorem 5.25. *Under assumption (5.4), if $T = D + F$ belongs to $\mathcal{D}^0_{per}(\mathbb{E}_\omega)$ where $De_k = \lambda_k e_k$ for all $k \in \mathbb{N}$ and $F = \sum_{k=1}^m u_k \otimes v_k$, then the spectrum $\sigma(T)$ of T is given by*

$$\sigma(T) = \left\{\lambda \in \mathbb{Q}_p \setminus \mathbb{Z}_p : \det M(\lambda) = 0\right\} \cup \mathbb{Z}_p.$$

Question 5.26. A interesting and open question consists of computing all the zeros of the function $\det M : \mathbb{Q}_p \setminus \mathbb{Z}_p \mapsto \mathbb{Q}_p$.

5.5 Bibliographical Notes

The material of this chapter (including proofs) is taken from the following sources: Diagana et al. [18], Diagana [13], and Diarra [19–21]. For more on the topic treated in this chapter in the classical setting, we refer the reader to Fang and Xia [25], Fois et al. [26], Gohberg et al. [27, 28], and Ionascu [29].

Chapter 6
Unbounded Linear Operators

This chapter introduces and studies unbounded operators on a non-archimedean Baanach space $\mathbb{X}$. Various properties of those operators will be discussed including their spectral theory. In this chapter, we mainly follow Diagana [13] and Diagana and Ramaroson [17].

6.1 Unbounded Linear Operators on a Non-archimedean Banach Space

Definition 6.1. An unbounded linear operator A on a non-archimedean Banach space $(\mathbb{X}, \| \cdot \|)$ is a pair $(\mathrm{Dom}(A), A)$ consisting of a subspace $\mathrm{Dom}(A) \subset \mathbb{X}$ (called the domain of A) and a (possibly not continuous) linear transformation $A : \mathrm{Dom}(A) \subset \mathbb{X} \mapsto \mathbb{X}$. The collection of all unbounded linear operators on $\mathbb{X}$ will be denoted $U(\mathbb{X})$.

The previous definition takes the following formulation in $\mathbb{E}_\omega$:

Definition 6.2. An unbounded linear operator A on $\mathbb{E}_\omega$ is a linear transformation $A : \mathrm{Dom}(A) \subset \mathbb{E}_\omega \mapsto \mathbb{E}_\omega$ whose domain $\mathrm{Dom}(A)$ contains the basis $(e_i)_{i \in \mathbb{N}}$ and consists of all $u = (u_i)_{i \in \mathbb{N}} \in \mathbb{E}_\omega$ such $Au = \sum_{i \in \mathbb{N}} u_i A e_i$ converges in $\mathbb{E}_\omega$, that is,

$$
\begin{cases}
\mathrm{Dom}(A) := \left\{ u = (u_i)_{i \in \mathbb{N}} \in \mathbb{E}_\omega : \lim_{i \to \infty} |u_i| \, \|A e_i\| = 0 \right\}, \\[2em]
Au = \left(\sum_{i,j \in \mathbb{N}} a_{ij} \, e'_j \otimes e_i \right) u \quad \text{for each } u \in \mathrm{Dom}(A).
\end{cases}
$$

© The Author(s) 2016
T. Diagana, F. Ramaroson, *Non-Archimedean Operator Theory*,
SpringerBriefs in Mathematics, DOI 10.1007/978-3-319-27323-5_6

It is easy to see that if A is a bounded linear operator, then $\mathrm{Dom}(A) = \mathbb{X}$. However, if $A \in U(\mathbb{X})$, then its domain $\mathrm{Dom}(A)$ does not in general coincide with $\mathbb{X}$ (see the next example, which was given by Diarra [21]).

Example 6.3. Suppose the non-archimedean field $(\mathbb{K}, |\cdot|)$ contains a square root of each of its elements. Let $\lambda = (\lambda_j)_{j \in \mathbb{N}}$ such that $\lambda_j \in \mathbb{K} \setminus \{0\}$ for each $j \in \mathbb{N}$, and

$$\limsup_{j \to \infty} \left| \lambda_j \right| = \infty.$$

Consider the linear operator on $\mathbb{E}_\omega$ defined by $Ae_j = \lambda_j e_j$ for all $j \in \mathbb{N}$ whose domain is

$$\mathrm{Dom}(A) = \left\{ u = (u_i)_{i \in \mathbb{N}} \in \mathbb{E}_\omega : \lim_{j \to \infty} \left| \lambda_j \right| \left\| u_j \right\| \left\| e_j \right\| = 0 \right\}.$$

In fact, $\mathrm{Dom}(A) \neq \mathbb{E}_\omega$. Indeed, choose a vector $u = (u_j)_{j \in \mathbb{N}}$ such that u_j satisfies $u_j^2 = \omega_j^{-1} \lambda_j^{-2}$ for all $j \in \mathbb{N}$. Using the assumption made on the field $\mathbb{K}$ it follows that $u_j \in \mathbb{K}$ for all $j \in \mathbb{N}$. Moreover, using the assumption on the sequence $\lambda = (\lambda_j)_{j \in \mathbb{N}}$, it follows that there exists a subsequence $(\lambda_{j_n})_{n \in \mathbb{N}}$ of $(\lambda_j)_{j \in \mathbb{N}}$ such that

$$\lim_{n \to \infty} \left| \lambda_{j_n} \right| = \infty.$$

Setting $\bar{u} = (u_{j_n})_{n \in \mathbb{N}}$, one obtains that $\bar{u} \in \mathbb{E}_\omega$. Indeed,

$$\lim_{n \to \infty} \left| u_{j_n} \right| \left\| e_{j_n} \right\| = \lim_{n \to \infty} \left| \lambda_{j_n} \right|^{-1} = 0.$$

Now

$$\lim_{n \to \infty} \left| u_{j_n} \right| \left| \lambda_{j_n} \right| \left\| e_{j_n} \right\| = 1 \neq 0$$

and hence $\bar{u} = (u_{j_n})_{n \in \mathbb{N}} \notin \mathrm{Dom}(A)$.

6.2 Closed Linear Operators

Let $A \in U(\mathbb{X})$. Define its graph $\mathcal{G}(A)$ by

$$\mathcal{G}(A) = \left\{ (x, Ax) \in \mathbb{X} \times \mathbb{X} : x \in \mathrm{Dom}(A) \right\}.$$

Recall that $\mathbb{X} \times \mathbb{X}$ is equipped with its natural topology defined by,

$$\left\|(x, y)\right\| = \max\left(\left\|x\right\|, \left\|y\right\|\right)$$

for all $(x, y) \in \mathbb{X} \times \mathbb{X}$, which makes it a Banach space.

Definition 6.4 ([21]). An operator $A \in U(\mathbb{X})$ is said to be closed if its graph $G(A)$, as a subset of $\mathbb{X} \times \mathbb{X}$, is closed. The operator A is said to be closable, if it has a closed extension. The collection of all elements of $U(\mathbb{X})$ which are closed will be denoted $C(\mathbb{X})$.

One should point out that the closedness of an operator $A \in U(\mathbb{X})$ can be characterized as follows: for all $(u_n)_{n \in \mathbb{N}} \subset \text{Dom}(A)$ such that $\|u_n - u\| \to 0$ and $\|Au_n - \xi\| \to 0$ as $n \to \infty$ for some $u \in \mathbb{X}$ and $\xi \in \mathbb{X}$, then $u \in \text{Dom}(A)$ and $Au = \xi$.

Remark 6.5. It is easy to see that every bounded linear operator A on $\mathbb{X}$ is closed. Similarly, if A is a bounded linear operator on $\mathbb{E}_\omega$ and if $B \in C(\mathbb{X})$, then their algebraic sum $S := A + B$ defined by

$$Su = Au + Bu \text{ for all } u \in \text{Dom}(S) = \text{Dom}(A) \cap \text{Dom}(B) = \text{Dom}(B)$$

is closed.

Example 6.6. Consider the linear operator D on $\mathbb{E}_\omega$ defined by $De_j = \lambda_j e_j$ for all $j \in \mathbb{N}$ and whose domain is

$$\text{Dom}(D) = \left\{ u = (u_j)_{j \in \mathbb{N}} \in \mathbb{E}_\omega : \lim_{j \to \infty} \left|\lambda_j\right| \left\|u_j\right\| \left\|e_j\right\| = 0 \right\}.$$

Explicitly, if $u \in \text{Dom}(D)$, one has $Du = \sum_{j \in \mathbb{N}} \lambda_j u_j e_j$.

Proposition 6.7. *The operator* $D : \text{Dom}(D) \subset \mathbb{E}_\omega \mapsto \mathbb{E}_\omega$ *defined above is closed.*

Proof. Let $(u_n)_{n \in \mathbb{N}} \in \text{Dom}(D)$ such that $u_n \to u$ and $Du_n \to v$ as $n \to \infty$ for some $u, v \in \mathbb{E}_\omega$.

Write

$$u_n = \sum_{j \in \mathbb{N}} a_j^n e_j, \quad u = \sum_{j \in \mathbb{N}} a_j e_j, \text{ and } v = \sum_{j \in \mathbb{N}} b_j e_j$$

where $a_j^n, a_j, b_j \in \mathbb{K}$ for all $j, n \in \mathbb{N}$ and $\lim_{j \to \infty} |a_j^n| \|e_j\| = 0$, $\lim_{j \to \infty} |a_j| \|e_j\| = 0$, and $\lim_{j \to \infty} |b_j| \|e_j\| = 0$ for all $n \in \mathbb{N}$.

Now from $u_n \to u$ and $Du_n \to v$ in $\mathbb{E}_\omega$ as $n \to \infty$ it follows that $|a_j^n - a_j| \to 0$ and $|\lambda_j a_j^n - b_j| \to 0$ as $n \to \infty$ for all $j \in \mathbb{N}$ which yields, $a_j \lambda_j = b_j$ for all $j \in \mathbb{N}$. Consequently, $u \in \text{Dom}(D)$ as

$$\lim_{j\to\infty}\left|\lambda_j\right|\left|a_j\right|\left\|e_j\right\| = \lim_{j\to\infty}\left|\lambda_j a_j\right|\left\|e_j\right\| = \lim_{j\to\infty}\left|b_j\right|\left\|e_j\right\| = 0.$$

Further, $v = Du$. Therefore, D is closed.

Example 6.8. Suppose $\omega_j = 1$ for all $j \in \mathbb{N}$ which yields $\|e_j\| = 1$. Let $V \subset \mathbb{E}_\omega$ be an infinite dimensional subspace and let $\xi : V \mapsto \mathbb{K}$ be an arbitrary linear functional. Suppose that the linear functional ξ is not continuous.

Consider the linear operator A on $\mathbb{E}_\omega$ defined by $Ax = \xi(x)e_1$ for all $x \in \mathrm{Dom}(A)$ where $\mathrm{Dom}(A) := V$.

Proposition 6.9. *The operator $A : V \subset \mathbb{E}_\omega \mapsto \mathbb{E}_\omega$ defined above is not closed.*

Proof. Using the fact ξ is not continuous it follows that there exists a sequence $(x'_n)_{n\in\mathbb{N}} \subset V$ such that $x'_n \to 0$ as $n \to \infty$ and $(\xi(x'_n))_{n\in\mathbb{N}}$ does not converge. We can assume that there exists $M > 0$ such that $|\xi(x'_n)| > M$ for all $n \in \mathbb{N}$. Setting $x_n = \xi(x'_n)^{-1}x'_n \in V$ it follows that $x_n \to 0$ as $n \to \infty$ while $Ax_n = e_1 \neq 0$. Therefore, A is not closed.

Let us point out that discontinuous linear functionals on $\mathbb{E}_\omega$ actually exist and can be constructed through the axiom of choice and Hamel bases. Indeed, using the axiom of choice, one can complete the canonical basis $\{e_j\}_{j\in\mathbb{N}}$ of $\mathbb{E}_\omega$ into a Hamel basis, which we denote by H_B. Now choosing a sequence $(\mu_j)_{j\in\mathbb{N}}$ such that $\mu_j \in \mathbb{K}$ for all $j \in \mathbb{N}$ and $\sup_{j\in\mathbb{N}} |\mu_j| = \infty$, one can see that the functional ξ defined by $\xi(e_j) = \mu_j$ for all $j \in \mathbb{N}$ and $\xi(e) = 0$ for all $e \in H_B \setminus \{e_j\}_{j\in\mathbb{N}}$, is a linear functional which is not continuous as

$$\|\xi\| = \sup_{j\in\mathbb{N}} \frac{\left|\xi(e_j)\right|}{\left\|e_j\right\|} = \sup_{j\in\mathbb{N}} \left|\mu_j\right| = \infty.$$

6.3 The Spectrum of an Unbounded Operator

Definition 6.10. The resolvent of an operator $A \in U(\mathbb{X})$ is defined by

$$\rho(A) := \left\{\lambda \in \mathbb{K} : \lambda I - A \text{ is a bijection, and } (\lambda I - A)^{-1} \in B(\mathbb{X})\right\}.$$

The spectrum $\sigma(A)$ of A is then defined by $\sigma(A) = \mathbb{K} \setminus \rho(A)$.

Definition 6.11. A scalar $\lambda \in \mathbb{K}$ is called an eigenvalue of $A \in U(\mathbb{X})$ whenever there exists a nonzero $u \in \mathrm{Dom}(A)$ (called eigenvector associated with λ) such that $Au = \lambda u$.

It is clear that eigenvalues of A consist of all $\lambda \in \mathbb{K}$ for which $\lambda I - A$ is not one-to-one, that is, $N(\lambda I - A) \neq \{0\}$. The collection of all eigenvalues is denoted $\sigma_p(A)$ (called point spectrum) and is defined by

$$\sigma_p(A) = \left\{ \lambda \in \sigma(A) : N(A - \lambda I) \neq \{0\} \right\}.$$

Definition 6.12. Define the essential spectrum $\sigma_e(A)$ of an unbounded linear operator $A : \mathrm{Dom}(A) \subset \mathbb{X} \mapsto \mathbb{X}$ as follows

$$\sigma_e(A) := \left\{ \lambda \in \mathbb{K} : \lambda I - A \text{ is not a Fredholm operator of index } 0 \right\}.$$

We have,

$$\sigma(A) = \sigma_p(A) \cup \sigma_e(A).$$

Note that the union $\sigma(A) = \sigma_p(A) \cup \sigma_e(A)$ is not a disjoint. It is easy to see that the intersection $\sigma_p(A) \cap \sigma_e(A)$ consists of eigenvalues λ of A for which,

(a) either $\dim N(\lambda I - A)$ is not finite;
(b) or $R(\lambda I - A)$ is not closed;
(c) or $\dim N(A) \neq \dim (\mathbb{X} \setminus R(A))$.

Definition 6.13. Define the continuous spectrum $\sigma_c(A)$ of an unbounded linear operator $A : \mathrm{Dom}(A) \subset \mathbb{X} \mapsto \mathbb{X}$ as follows

$$\sigma_c(A) := \left\{ \lambda \in \sigma_e(A) \setminus \sigma_p(A) : \overline{R(\lambda I - A)} = \mathbb{X} \right\}.$$

Definition 6.14. Define the residual spectrum $\sigma_r(A)$ of an unbounded linear operator $A : \mathrm{Dom}(A) \subset \mathbb{X} \mapsto \mathbb{X}$ as follows

$$\sigma_r(A) := \left(\sigma_e(A) \setminus \sigma_p(A) \right) \setminus \sigma_c(A).$$

We have

$$\sigma(A) = \sigma_p(A) \cup \sigma_c(A) \cup \sigma_r(A).$$

6.4 Unbounded Fredholm Operators

Definition 6.15. An operator $A \in U(\mathbb{X})$ is said to be a Fredholm operator if, A is closed, and if the integers $\eta(A) := \dim N(A)$ and $\delta(A) := \dim(\mathbb{X} \setminus R(A))$ are finite.

Note that if A is a Fredholm operator, then $R(A)$ is closed. The collection of all (possibly unbounded) Fredholm linear operators on $\mathbb{X}$ is denoted by $\Phi(\mathbb{X})$.

If $A \in \Phi(\mathbb{X})$, we then define its index by setting,

$$\chi(A) := \eta(A) - \delta(A).$$

Classical examples of Fredholm operators include invertible operators.

Let $\lambda = (\lambda_j)_{j \in \mathbb{N}}$ be a sequence such that $\lambda_j \in \mathbb{K}$ for each $j \in \mathbb{N}$ and such that

$$0 < \liminf_{j \to \infty} \left| \lambda_j \right| \neq \limsup_{j \to \infty} \left| \lambda_j \right| = \infty. \tag{6.1}$$

Example 6.16. Consider the unbounded diagonal linear operator D given in Example 6.6, that is, $De_j = \lambda_j e_j$ for all $j \in \mathbb{N}$ and whose domain is

$$\mathrm{Dom}(D) = \left\{ u = (u_j)_{j \in \mathbb{N}} \in \mathbb{E}_\omega : \lim_{j \to \infty} \left| \lambda_j \right| \left\| u_j \right\| \left\| e_j \right\| = 0 \right\}.$$

Proposition 6.17. *The operator $D : \mathrm{Dom}(D) \subset \mathbb{E}_\omega \mapsto \mathbb{E}_\omega$ defined above is a Fredholm operator with $\chi(D) = 0$.*

Proof. We have already shown that D is a closed linear operator (see Proposition 6.7). Now using Eq. (6.1) it follows that $d = \#\{\lambda_j : \lambda_j = 0\}$ is finite. Consequently, $\dim N(D) = d = \dim(\mathbb{E}_\omega \setminus R(D))$. Hence, $D \in \Phi(\mathbb{E}_\omega)$ with $\chi(D) = 0$.

Theorem 6.18. *If $A \in \Phi(\mathbb{E}_\omega)$, then for all $K \in C(\mathbb{E}_\omega)$, we have $A + K \in \Phi(\mathbb{E}_\omega)$ with*

$$\chi(A + K) = \chi(A).$$

Proof. Since $A \in \Phi(\mathbb{E}_\omega)$, then A is closed. Consequently, $A + K$ is closed as the algebraic sum of a closed and a bounded linear operators (see Remark 6.5). Let $\widetilde{\mathrm{Dom}}(A)$ denote the normed vector space $(\mathrm{Dom}(A), \| \cdot \|_{\mathrm{Dom}(A)})$, where $\| \cdot \|_{\mathrm{Dom}(A)}$ is the so-called non-archimedean graph norm defined by

$$\left\| x \right\|_{\mathrm{Dom}(A)} = \max \left(\left\| x \right\|, \left\| Ax \right\| \right)$$

for all $x \in \mathrm{Dom}(A)$.

Since A is a closed linear operator, then $\widetilde{\mathrm{Dom}}(A)$ is a non-archimedean Banach space. We now regard A and K (the restriction of K to $\mathrm{Dom}(A)$) as linear operators from $\widetilde{\mathrm{Dom}}(A)$ to $\mathbb{E}_\omega$. These operators will be denoted respectively by $\tilde{A}$ and $\tilde{K}$. It is easy to see that both $\tilde{A}$ and $\tilde{K}$ are bounded linear operators from $\widetilde{\mathrm{Dom}}(A)$ to $\mathbb{E}_\omega$. It is also clear that $\tilde{K}$ is a completely continuous linear operator.

Now $R(A) = R(\tilde{A})$ and $R(A + K) = R(\tilde{A} + \tilde{K})$. Further, we have $\eta(A) = \eta(\tilde{A})$, $\delta(A) = \delta(\tilde{A})$, $\eta(A+K) = \eta(\tilde{A}+\tilde{K})$, and $\delta(A+K) = \delta(\tilde{A}+\tilde{K})$. Consequently, $\tilde{A}$ is a Fredholm operator. Using Theorem 5.1 on the sum of bounded Fredholm operators it follows that $\tilde{A} + \tilde{K} : \widetilde{\mathrm{Dom}}(A) \mapsto \mathbb{E}_\omega$ is a Fredholm operator with index,

$$\chi(\tilde{A} + \tilde{K}) = \chi(\tilde{A}) = \chi(A).$$

Now using the facts that $\delta(A+K) = \delta(\tilde{A}+\tilde{K}) < \infty$ and $\eta(A+K) = \eta(\tilde{A}+\tilde{K}) < \infty$ it follows that $A + K$ is a Fredholm operator with index,

$$\chi(A + K) = \chi(\tilde{A} + \tilde{K}) = \chi(\tilde{A}) = \chi(A).$$

6.5 Bibliographical Notes

The material of this chapter is taken from the following sources: Diagana [13, 16] and Diagana and Ramaroson [17].

Chapter 7
Spectral Theory for Perturbations of Unbounded Linear Operators

In this chapter we first study the spectral theory of completely continuous perturbations of unbounded Fredholm operators in the non-archimedean Hilbert space $\mathbb{E}_\omega$. Next, we make extensive use of these results to compute the spectrum of the class of linear operators on $\mathbb{E}_\omega$ of the form

$$A = D + F,$$

where D is an unbounded diagonal operator and F is a finite rank operator.

7.1 Introduction

Recall that for a given sequence $(\lambda_j)_{j \in \mathbb{N}}$ with $\lambda_j \in \mathbb{K}$ for all $j \in \mathbb{N}$, we set $\Lambda = \{\lambda_j : j \in \mathbb{N}\}$. Similarly, for each $\lambda \in \Lambda$, we let $I_\lambda = \{k \in \mathbb{N} : \lambda_k = \lambda\}$ and let $r_\lambda = \#I_\lambda$, where # denotes the cardinal of a set.

Set

$$\Lambda^* = \left\{ \lambda \in \Lambda : r_\lambda < \infty \right\}$$

and

$$\Lambda' = \left\{ \lambda \in \Lambda : \lambda \text{ is an accumulation point of } \Lambda \right\}.$$

Let $\Lambda^\infty = \{\lambda \in \Lambda : r_\lambda = \infty\}$ and let $\partial\Lambda = \overline{\Lambda} \setminus \Lambda$ be the boundary of Λ in $\mathbb{K}$. The main objective of this chapter consists of extending the results of Chap. 5 to the case of unbounded linear operators on $\mathbb{E}_\omega$. More precisely, we show that if

© The Author(s) 2016
T. Diagana, F. Ramaroson, *Non-Archimedean Operator Theory*,
SpringerBriefs in Mathematics, DOI 10.1007/978-3-319-27323-5_7

$$A = D + F$$

where D is an unbounded diagonal operator and $F = u_1 \otimes v_1 + u_2 \otimes v_2 + \ldots + u_m \otimes v_m$ is a finite rank operator, then

$$\sigma(A) = \sigma_e(D) \cup \sigma_p(F),$$

where

$$\sigma_p(A) = \left\{ \lambda \in \rho(D) : \det M(\lambda) = 0 \right\},$$

and

$$\sigma_e(D) = \partial \Lambda \cup \Lambda^\infty \cup \Lambda'.$$

7.2 Spectral Analysis for the Class of Operators $T = D + K$

Let $\lambda = (\lambda_j)_{j \in \mathbb{N}}$ be a sequence such that $\lambda_j \in \mathbb{K}$ for each $j \in \mathbb{N}$. Further, suppose that $\lambda = (\lambda_j)_{j \in \mathbb{N}}$ satisfies Eq. (6.1).

Consider the linear operator on $\mathbb{E}_\omega$ defined by $De_j = \lambda_j e_j$ for all $j \in \mathbb{N}$ whose domain is

$$\text{Dom}(D) = \left\{ u = (u_i)_{i \in \mathbb{N}} \in \mathbb{E}_\omega : \lim_{j \to \infty} \left| \lambda_j \right| \left\| u_j \right\| \left\| e_j \right\| = 0 \right\}.$$

This section is devoted to the study of the spectral analysis for perturbations of the unbounded diagonal operator D by completely continuous operators. More precisely, we study the spectral theory of the class of linear operators of the form

$$T = D + K,$$

where $K : \mathbb{E}_\omega \mapsto \mathbb{E}_\omega$ is a completely continuous linear operator.

An immediate consequence of Theorem 6.18 is given by:

Corollary 7.1. *If $A \in \Phi(\mathbb{E}_\omega)$, then for all $K \in C(\mathbb{E}_\omega)$, we have $\sigma_e(A + K) = \sigma_e(A)$.*

Proof. If λ does not belong to $\sigma_e(A)$, then $\lambda I - A$ belongs to $\Phi(\mathbb{E}_\omega)$ with $\chi(\lambda I - A) = 0$. Using Theorem 6.18 it follows that $\lambda I - A - K$ belongs to $\Phi(\mathbb{E}_\omega)$ with $\chi(\lambda I - A - K) = 0$ for all $K \in C(\mathbb{E}_\omega)$.

Corollary 7.2. *For every $K \in C(\mathbb{E}_\omega)$, we have $\sigma_e(D + K) = \sigma_e(D)$.*

From Corollary 7.2 and the fact that $\sigma(T) = \sigma_p(T) \cup \sigma_e(T)$, we obtain the following important result.

Proposition 7.3. *If $T = D + K$ where D is an unbounded diagonal operator whose coefficients satisfy Eq. (6.1) and $K \in C(\mathbb{E}_\omega)$, then its spectrum $\sigma(T)$ is given by $\sigma(T) = \sigma_e(D) \cup \sigma_p(T)$.*

Corollary 7.4. *If $T = D + u_1 \otimes v_1 + u_2 \otimes v_2 + \ldots + u_m \otimes v_m$, where D is an unbounded diagonal operator whose coefficients satisfy Eq. (6.1), then its spectrum $\sigma(T)$ is given by $\sigma(T) = \sigma_e(D) \cup \sigma_p(T)$.*

7.3   Spectral Analysis for the Class of Operators $T = D + F$

In this section we study the spectral theory of linear operators of the form

$$T = D + F$$

where $D : \mathrm{Dom}(D) \subset \mathbb{E}_\omega \mapsto \mathbb{E}_\omega$ is an unbounded diagonal operator defined by $De_j = \lambda_j e_j$, with the sequence $\lambda = (\lambda_j)_{j \in \mathbb{N}}$ satisfying Eq. (6.1), and F is an operator of finite rank defined by

$$F = \sum_{k=1}^{m} u_k \otimes v_k,$$

where $u_k = (\alpha_j^k)_{j \in \mathbb{N}}$, $v_k = (\beta_j^k)_{j \in \mathbb{N}} \in \mathbb{E}_\omega$ with $\alpha_j^k, \beta_j^k \in \mathbb{K} \setminus \{0\}$ for each $k = 1, 2, \ldots, m$ and $j \in \mathbb{N}$.

Let $\mathcal{D}_U^0(\mathbb{E}_\omega)$ denote the collection of all unbounded linear operators of the form

$$T = D + F = D + u_1 \otimes v_1 + u_2 \otimes v_2 + \ldots + u_m \otimes v_m,$$

where $D : \mathrm{Dom}(D) \subset \mathbb{E}_\omega \mapsto \mathbb{E}_\omega$ is an unbounded diagonal operator defined by $De_j = \lambda_j e_j$ and such $u_k = (\alpha_j^k)_{j \in \mathbb{N}}$, $v_k = (\beta_j^k)_{j \in \mathbb{N}} \in \mathbb{E}_\omega$ with $\alpha_j^k, \beta_j^k \in \mathbb{K} \setminus \{0\}$ for each $k = 1, 2, \ldots, m$ and $j \in \mathbb{N}$.

In the rest of this section, we suppose Eq. (5.4) holds, that is, that there exists $d_k \neq 0$ for $k = 1, 2, \ldots, m$ such that

$$\langle u_k, u_l \rangle = d_k \delta_{kl} \tag{7.1}$$

for $k, l = 1, 2, \ldots, m$.

Lemma 7.5. *If $T = D + F$ belongs to $\mathcal{D}_U^0(\mathbb{E}_\omega)$, then $\lambda \in \sigma_p(T)$ if and only if*

(a) $\lambda \notin \sigma_p(D) = \left\{ \lambda_j : j \in \mathbb{N} \right\}$; i.e., $\lambda \neq \lambda_j$ for all $j \in \mathbb{N}$; and
(b) $\det M(\lambda) = 0$ where $M(\lambda) = (c_{ij}(\lambda))$ and $c_{ij}(\lambda) = \delta_{ij} - \langle C_\lambda u_j, v_i \rangle$ for $i, j = 1, 2, \ldots, m$ with $C_\lambda := (\lambda I - D)^{-1}$.

Proof. The proof follows along the same lines as that of Lemma 5.9. But for the sake of clarity, we reproduce it here with some slight modifications. Suppose $\lambda \in \sigma_p(T)$ and thus there exists $0 \neq w \in \mathrm{Dom}(T) = \mathrm{Dom}(D)$ such that $Tw = \lambda w$, that is,

$$(\lambda I - D)w = Fw = \sum_{k=1}^{m} \langle v_k, w \rangle u_k. \tag{7.2}$$

Clearly, not all the expressions $\langle v_k, w \rangle$ are zero. If not, we will get $(\lambda I - D)w = 0$ with $w \neq 0$, which yields $\lambda \in \sigma_p(D)$. Consequently, there exists $j_0 \in \mathbb{N}$ such that $\lambda = \lambda_{j_0}$, $w = ae_{j_0} \in \mathrm{Dom}(D)$ with $a \in \mathbb{K} \setminus \{0\}$ and $\omega_{j_0} \beta_{j_0}^k = \langle v_k, e_{j_0} \rangle = a^{-1}\langle v_k, w \rangle = 0$ yields $\beta_{j_0}^k = 0$ for $k = 1, 2, \ldots, m$, which contradicts the fact that $\beta_j^k \in \mathbb{K} \setminus \{0\}$ for each $k = 1, 2, \ldots, m$ and $j \in \mathbb{N}$. Consequently, $Fw = (\lambda I - D)w \neq 0$ and hence $u_k \in R(\lambda I - D)$ for $k = 1, 2, \ldots, m$ and $\lambda \notin \sigma_p(D)$.

Since $\lambda \in \rho(D)$, Eq. (7.2) is equivalent to

$$w - \sum_{k=1}^{m} \langle v_k, w \rangle C_\lambda u_k = 0. \tag{7.3}$$

Taking the inner product of Eq. (7.3) with respectively $v_1, v_2, \ldots,$ and v_m, then we obtain the following system of equations

$$M(\lambda) \begin{pmatrix} \langle w, v_1 \rangle \\ \langle w, v_2 \rangle \\ . \\ . \\ \langle w, v_m \rangle \end{pmatrix} = \begin{pmatrix} 0 \\ 0 \\ . \\ . \\ 0 \end{pmatrix}. \tag{7.4}$$

Using the fact that at least one of the following numbers $\langle w, v_1 \rangle, \langle w, v_2 \rangle, \ldots, \langle w, v_m \rangle$ is nonzero it follows that Eq. (7.4) has at least one nontrivial solution which yields $\det M(\lambda) = 0$.

Suppose (a)–(b) are true, that is, $\lambda \notin \sigma_p(D)$ and $\det M(\lambda) = 0$. Then there exist some scalars $\xi_1, \ldots, \xi_m$ not all zeros, such that with $\xi = (\xi_1, \ldots, \xi_m)^t$ we have

$$M(\lambda) \begin{pmatrix} \xi_1 \\ \xi_2 \\ . \\ . \\ \xi_m \end{pmatrix} = \begin{pmatrix} 0 \\ 0 \\ . \\ . \\ 0 \end{pmatrix}. \tag{7.5}$$

Guided by Eq. (7.3), we take $w = \sum_{k=1}^{m} \xi_k C_\lambda u_k \in \mathrm{Dom}(D)$ and obtain $(T - \lambda I)w = 0$. Now $w \neq 0$. For that, let us show that $C_\lambda^{-1} w \neq 0$ which, by using C_λ, yields $w \neq 0$. If $C_\lambda^{-1} w = 0$ it follows that $0 = \langle C_\lambda^{-1} w, u_j \rangle = -\xi_j d_j$ for $j = 1, 2, \ldots, m$

which yields $\xi_j = 0$ for $j = 1, 2, \ldots, m$ and that contradicts the fact that some of the ξ_j are nonzero. In view of the above, $N(T - \lambda I) \neq 0$. That is, $\lambda \in \sigma_p(T)$.

We also have:

Corollary 7.6. *Let* $T = D + F$ *belong to* $\mathcal{D}_U^0(\mathbb{E}_\omega)$ *and let* $\lambda \in \rho(D)$. *Then,* $\lambda \in \sigma_p(T)$ *if and only if* $\det M(\lambda) = 0$ *where* $M(\lambda) = (c_{ij}(\lambda))$ *with* $c_{ij}(\lambda) = \delta_{ij} - \langle C_\lambda u_j, v_i \rangle$ *for* $i, j = 1, 2, \ldots, m$.

Corollary 7.7. *If* $T = D + F$ *belongs to* $\mathcal{D}_U^0(\mathbb{E}_\omega)$ *where* $F = \sum_{k=1}^m u_k \otimes v_k$, *then the eigenvalues of* T *are given by*

$$\sigma_p(T) = \left\{ \lambda \in \rho(D) : \det M(\lambda) = 0 \right\}.$$

Corollary 7.8. *If* $T = D + F$ *belongs to* $\mathcal{D}_U^0(\mathbb{E}_\omega)$ *where* $F = \sum_{k=1}^m u_k \otimes v_k$, *then the spectrum* $\sigma(T)$ *of* T *is given by*

$$\sigma(T) = \left\{ \lambda \in \rho(D) : \det M(\lambda) = 0 \right\} \cup \sigma_e(D).$$

7.4 Computation of $\sigma_e(D)$

Our main objective in this section consists of determining, in as concrete a way as possible, the essential spectrum of an *unbounded* diagonal operator D without any further conditions on the diagonal entries.

We set the following notations:

- $\Lambda = \{\lambda_j \in \mathbb{K} : j \in \mathbb{N}\}$ is the set of diagonal entries.
- $\overline{\Lambda}$ is the closure of Λ in $\mathbb{K}$.
- $\Lambda' = \{\lambda \in \Lambda : \lambda$ is an accumulation point of $\Lambda\}$.
- For each λ in Λ, $I_\lambda = \{j \in \mathbb{N} : \lambda_j = \lambda\}$. Further, $r_\lambda = \#(I_\lambda) =$ cardinality of I_λ.
- $\Lambda^* = \{\lambda \in \Lambda : r_\lambda < \infty\}$.
- $\Lambda^\infty = \{\lambda \in \Lambda : r_\lambda = \infty\}$.
- $\partial\Lambda = \overline{\Lambda} \setminus \Lambda$, the boundary of Λ in $\mathbb{K}$.

Recall that Dom (D) the domain of D is given by

$$\text{Dom}\,(D) = \left\{ u = (u_j)_{j \in \mathbb{N}} \in \mathbb{E}_\omega : \lim_{j \to \infty} |\lambda_j|\, |u_j|\, \|e_j\| = 0 \right\}$$

and for each u in Dom (D)

$$Du = \sum_{j \in \mathbb{N}} \lambda_j u_j e_j.$$

We begin with the following Proposition, which is a slight refinement of [20, Proposition 4.1.].

Proposition 7.9. *Let D be an unbounded diagonal operator as above, then, D is self-adjoint and* $\rho(D) = \mathbb{K} \setminus \overline{\Lambda}$.

Proof. Clearly, $D = \sum\limits_{i,j \in \mathbb{N}} a_{ij}(e_i' \otimes e_j)$ where $a_{ij} = \lambda_i \delta_{ij}$ and δ_{ij} is Kronecker's symbol.

As in [20], the diagonal operator D is well-defined, moreover,

$$\text{for every } j, \lim_{i \to \infty} \left| a_{ij} \right| \left\| e_i \right\| = \lim_{i > j} \left| \lambda_i \delta_{ij} \right| \left\| e_i \right\| = 0$$

and

$$\text{for every } i, \lim_{j \to \infty} \frac{\left| a_{ij} \right|}{\left| \omega_j \right|^{1/2}} = \lim_{j > i} \frac{\left| \lambda_i \delta_{ij} \right|}{\left| \omega_j \right|^{1/2}} = 0,$$

therefore D has an adjoint. As the adjoint is

$$D^* = \sum_{i,j \in \mathbb{N}} a_{ij}^*(e_i' \otimes e_j) \text{ with } a_{ij}^* = \omega_i^{-1} \omega_j a_{ji}.$$

We easily see that $a_{ij}^* = a_{ij}$ and hence $D^* = D$ and D is self-adjoint.

Let $\lambda \in \rho(D)$ and suppose that $\lambda \in \overline{\Lambda}$. There exists $\{\lambda_{k_j}\} \subset \Lambda$ such that $\lim_j \lambda_{k_j} = \lambda$. In other words, for every $\varepsilon > 0$ there exists J such that for $j \geq J$,

$$\left| \lambda_{k_j} - \lambda \right| < \varepsilon.$$

Let

$$\varepsilon = \frac{1}{\left\| (D - \lambda I)^{-1} \right\|} < \infty \quad (\lambda \in \rho(D)).$$

Then there exists J such that for $j \geq J$,

$$\left| \lambda_{k_j} - \lambda \right| < \frac{1}{\left\| (D - \lambda I)^{-1} \right\|}$$

and hence for all $j \geq J$,

$$\left\| (D - \lambda I)^{-1} \right\| \geq \frac{\left\| (D - \lambda I)^{-1} \left(e_{k_j} \right) \right\|}{\left\| e_{k_j} \right\|} = \frac{1}{\left| \lambda_{k_j} - \lambda \right|} > \left\| (D - \lambda I)^{-1} \right\|.$$

This contradiction implies that λ lies in $\mathbb{K} \setminus \overline{\Lambda}$.

Next, let $\lambda \in \mathbb{K} \setminus \overline{\Lambda}$. This implies that there exists $\epsilon > 0$ such that for every i, $|\lambda_i - \lambda| \geq \epsilon$. Following [20], we need to consider, for each i, the quantities $a_i = |\lambda_i - \lambda|^{-1}$ and $b_i = |\lambda_i| \cdot |\lambda_i - \lambda|^{-1}$ and we want to show that they are bounded, in other words, there exists c_λ such that $a_i \leq c_\lambda$ and $b_i \leq c_\lambda$ for every i.

These will imply, as in [20], that

(1) for every $y \in \mathbb{E}_\omega$ the equation $(D - \lambda I) x = y$ has a unique solution x in Dom (D), and hence $D - \lambda I$ is bijective.
(2) $(D - \lambda I)^{-1} \in B(\mathbb{E}_\omega)$.

These two facts now imply that λ lies in $\rho(D)$.

Lemma 7.10. *The quantities a_i and b_i are bounded.*

Proof. Consider the following sets,

(a) $\{i : |\lambda_i| < |\lambda|\}$;
(b) $\{i : |\lambda_i| > |\lambda|\}$; and
(c) $\{i : |\lambda_i| = |\lambda|\}$.

If i is in case (a), then $|\lambda_i - \lambda| = |\lambda|$

$$a_i = \frac{1}{\left| \lambda \right|},$$

$$b_i = \frac{\left| \lambda_i \right|}{\left| \lambda \right|} < 1.$$

If i is in case (b), then $|\lambda_i - \lambda| = |\lambda_i|$

$$a_i = \frac{1}{\left| \lambda_i \right|} < \frac{1}{\left| \lambda \right|},$$

$$b_i = 1.$$

If i is in case (c), then we use $|\lambda_i - \lambda| \geq \epsilon$ to obtain,

$$a_i \leq \frac{1}{\epsilon},$$

$$b_i \leq \frac{|\lambda_i|}{\varepsilon} = \frac{|\lambda|}{\varepsilon}.$$

Now letting

$$c_\lambda = \max\left(1, \frac{1}{\varepsilon}, \frac{1}{|\lambda|}, \frac{|\lambda|}{\varepsilon}\right).$$

we obtain that, for every i, a_i and b_i are bounded above by c_λ.

Corollary 7.11. $\sigma(D) = \overline{\Lambda}$.

For each λ in Λ, let V_λ be the subspace of $\mathbb{E}_\omega$ defined by

$$V_\lambda = \left\{\sum_i v_i e_i : v_i = 0 \text{ for each } i \text{ in } I_\lambda\right\}.$$

Then V_λ is a closed subspace of $\mathbb{E}_\omega$.

Theorem 7.12. *With the notation given in the beginning of this section, we have*

$$\sigma_e(D) = \partial\Lambda \cup \Lambda^\infty \cup \Lambda'.$$

Proof. First of all, let us note that $\sigma(D) = \overline{\Lambda} = \sigma_p(D) \cup \sigma_e(D) = \Lambda \cup \sigma_e(D)$. Let $\lambda \in \sigma_e(D)$, then $\lambda \in \overline{\Lambda}$·and if $\lambda \notin \Lambda$ then $\lambda \in \partial\Lambda$. If $\lambda \in \Lambda$ and $r_\lambda = \infty$ then $\lambda \in \Lambda^\infty$. So suppose $\lambda \in \Lambda^*$ and assume that $\lambda \notin \Lambda'$. The operator $\lambda I - D$ is closed. Further, it can be shown that $\lambda I - D \in \Phi_0(\mathbb{E}_\omega)$ contradicting the fact that $\lambda \in \sigma_e(D)$. Hence $\lambda \in \Lambda'$. We conclude that $\sigma_e(D) \subset \partial\Lambda \cup \Lambda^\infty \cup \Lambda'$.

For the inclusion in the other direction, we first observe that $\partial\Lambda \cup \Lambda^\infty$ is clearly contained in $\sigma_e(D)$. So let $\lambda \in \Lambda'$. Now using Lemma 5.18, Corollary 5.19, and Lemma 5.20 we see that $\lambda \in \sigma_e(D)$. The proof is complete.

Corollary 7.13. *Suppose* $\lim_{j\to\infty} |\lambda_j| = \infty$, *then* $\sigma_e(D) = \varnothing$.

Proof. In this case, $\overline{\Lambda} = \Lambda$, $\Lambda^\infty = \Lambda' = \varnothing$.

7.5 Main Result

Using Corollary 7.8 and Theorem 7.12, we obtain:

Theorem 7.14. *If* $T = D + F$ *belongs to* $\mathcal{D}_U^0(\mathbb{E}_\omega)$ *and if Eq. (6.1) holds, then the spectrum* $\sigma(T)$ *of* T *is given by*

$$\sigma\left(T\right) = \{\lambda \in \rho\left(D\right) : \det M\left(\lambda\right) = 0\} \cup \partial\Lambda \cup \Lambda^{\infty} \cup \Lambda',$$

where $M\left(.\right)$ is the $m \times m$ square matrix given in Lemma 7.5.

7.6 Bibliographical Notes

The material of this chapter is taken from the following sources: Diagana et al. [18], Diagana [13, 16], Diarra [19–21], and Diagana and Ramaroson [17].

Appendix A
The Shnirel'man Integral

This section is devoted to the properties of the so-called Shnirel'man Integral which will be needed to study the functional calculus of bounded linear operators on $\mathbb{E}_\omega$.

Definition A.1. Let $\sigma \subset \mathbb{K}$ be a subset and let $r > 0$. The sets $D(\sigma, r)$ and $D(\sigma, r^-)$ are defined respectively as follows:

$$D(\sigma, r) := \left\{ x \in \mathbb{K} : dist(x, \sigma) \leq r \right\}$$

and

$$D(\sigma, r^-) := \left\{ x \in \mathbb{K} : dist(x, \sigma) < r \right\},$$

where $dist(x, \sigma) = \inf_{y \in \sigma} |x - y|$.

Additionally, for $a \in \mathbb{K}$, we define $D(a, r^-)$ and $D(a, r)$ respectively by

$$D(a, r^-) := \left\{ x \in \mathbb{K} : |x - a| < r \right\}$$

and

$$D(a, r) := \left\{ x \in \mathbb{K} : |x - a| \leq r \right\}.$$

Lemma A.2 ([7]). *Let $\sigma \subset \mathbb{K}$ be a nonempty compact subset. Then for every $s > 0$, there exist $0 < r \in |\mathbb{K}|$ and $a_1, \cdots, a_N \in \sigma$ such that $r < s$ and*

© The Author(s) 2016
T. Diagana, F. Ramaroson, *Non-Archimedean Operator Theory*,
SpringerBriefs in Mathematics, DOI 10.1007/978-3-319-27323-5

$$D(\sigma, r) = \bigsqcup_{i=1}^{N} D(a_i, r) \ \text{ and } \ \sigma \subset \bigsqcup_{i=1}^{N} D(a_i, r^-),$$

where the symbol $\bigsqcup$ denotes disjoint unions.

One can generalize Lemma 1.3 in Baker [7] as follows:

Lemma A.3. *Let* $\emptyset \neq \sigma \subset \mathbb{K}$ *and let* $r > 0$. *Then if* I *is a nonempty set and if* $\{b_i : i \in I\} \subset \mathbb{K}$ *is a subset such that*

$$\sigma \subset \bigsqcup_{i \in I} D(b_i, r^-),$$

then there exist a subset $J \subset I$ *and subset* $\{a_j : j \in J\} \subset \sigma$ *such that*

$$D(\sigma, r^-) = \bigsqcup_{j \in J} D(a_j, r^-) = \bigsqcup_{j \in L} D(b_j, r^-), \ \text{ and}$$

$$D(\sigma, r) = \bigsqcup_{j \in L} D(a_j, r) = \bigsqcup_{j \in J} D(b_j, r).$$

Proof. Set $J = \{j \in I : D(b_j, r^-) \cap \sigma \neq \emptyset\}$ and rewrite J as $J = \{i_j : j \in J\}$. For all $j \in J$, choose $a_j \in D(b_{i_j}, r^-)$. Then, $D(b_{i_j}, r^-) = D(a_j, r^-)$ and thus,

$$\sigma \subset \bigsqcup_{j \in J} D(a_j, r^-) = \bigsqcup_{j \in J} D(b_{i_j}, r^-) = \bigsqcup_{j \in J} D(b_j, r^-).$$

Obviously,

(i) $\bigsqcup_{j \in J} D(a_j, r^-) \subset D(\sigma, r^-)$;

(ii) $\bigsqcup_{j \in J} D(a_j, r) \subset D(\sigma, r)$.

Now we show the reverse inclusions. Let $x \in D(\sigma, r^-)$. Then $dist(x, \sigma) < r$ and hence, there exists $a \in \sigma$ such that $|x - a| < r$. Since $a \in \sigma = \bigsqcup_{j \in J} D(a_j, r^-)$, there exists $a_{j_0} \in \sigma$ with $j_0 \in J$ such that $|a - a_{j_0}| < r$.

Now

$$|x - a_{j_0}| \leq \max\{|x - a|, |a - a_{j_0}|\} < r,$$

and hence $x \in \bigsqcup_{j \in J} D(a_j, r^-)$. Therefore, $D(\sigma, r^-) \subset \bigsqcup_{j \in J} D(a_j, r^-)$.

Finally, let $x \in D(\sigma, r)$, that is, $dist(x, \sigma) \le r$ and there exists $a \in \sigma$ such that $|x - a| \le r$. Again, since $a \in \sigma$, there exists $a_{j_1} \in \sigma$ such that $|a - a_{j_1}| < r$. Now

$$|x - a_{j_1}| \le \max\{|x - a|, |a - a_{j_1}|\} \le r,$$

and hence $x \in \bigsqcup_{j \in J} D(a_j, r)$ and therefore, $D(\sigma, r) \subset \bigsqcup_{j \in J} D(a_j, r)$.

The following Corollary is then immediate and hence its proof is omitted.

Corollary A.4. *Let $\emptyset \ne \sigma \subset \mathbb{K}$ and let $r > 0$. Let $b_1, \cdots, b_M$ be in $\mathbb{K}$ with*

$$\sigma \subset \bigsqcup_{i=1}^{M} D(b_i, r^-).$$

Then there exist $a_1, \cdots, a_N$ in σ and $\emptyset \ne J \subset \{1, \cdots, M\}$ such that the $D(a_i, r)$ are disjoint and

$$D(\sigma, r^-) = \bigsqcup_{i=1}^{N} D(a_i, r^-) = \bigsqcup_{i \in J} D(b_i, r^-)$$

and

$$D(\sigma, r) = \bigsqcup_{i=1}^{N} D(a_i, r) = \bigsqcup_{i \in J} D(b_i, r).$$

The notion of (local) analyticity in the next definition plays a crucial role throughout the paper.

Definition A.5. Let $a \in \mathbb{K}$ and let $r > 0$. A function $f : D(a, r) \mapsto \mathbb{K}$ is said to be analytic if f can be represented by a power series on $D(a, r)$, that is,

$$f(x) = \sum_{k=0}^{\infty} c_k(x - a)^k \quad \text{with} \quad \lim_{k \to \infty} r^k |c_k| = 0.$$

Remark A.6. Let $a \in \mathbb{K}$ and let $r > 0$. The function $f : D(a, r) \mapsto \mathbb{K}$ is said to be 'Krasner analytic' if it is a uniform limit of rational functions with poles belong to the complement of $D(a, r)$. In fact if $r \in |\mathbb{K}|$, it can be shown that a function analytic over $D(a, r)$ in the sense of Krasner is also analytic in the sense of Definition 4.4, see, e.g., [32].

Definition A.7. Let $\emptyset \ne \sigma \subset \mathbb{K}$ and let $r > 0$. Let $B_r(\sigma)$ be the collection of all functions $f : D(\sigma, r) \to \mathbb{K}$ such that f is analytic on $D(a, r)$ whenever $a \in \mathbb{K}$ and $D(a, r) \subset D(\sigma, r)$. If f is bounded on $D(\sigma, r)$, we then set

$$\left\| f \right\|_r = \max_{x \in D(\sigma, r)} \left| f(x) \right|.$$

One should point out that the notion of local analyticity appearing in Definition A.8 is new and due to Baker [7]. Additional comments on this new notion can be found in Remark A.10.

Definition A.8 ($L(\sigma)$). Let $\emptyset \neq \sigma \subset \mathbb{K}$. Define $L(\sigma)$ to be the collection of all $\mathbb{K}$-valued functions f for which there exist $a_1, \cdots, a_N$ in $\mathbb{K}$ and $0 < r \in \left| \mathbb{K} \right|$ such that

$$\sigma \subset \bigsqcup_{i=1}^{N} D(a_i, r^-),$$

where the $D(a_i, r)$ are disjoint and f is analytic on each $D(a_i, r)$.

The class of functions $L(\sigma)$ will be called as the set of locally analytic functions on σ. Note that in view of Definition A.8, $Dom(f)$, the domain of $f \in L(\sigma)$ is

$$Dom(f) \subset \bigsqcup_{i=1}^{N} D(a_i, r).$$

Moreover, $L(\sigma) \neq \emptyset$, as polynomials belong to it.

Theorem A.9 ([7]). *Let $\emptyset \neq \sigma \subset \mathbb{K}$ be a compact subset. Then*

$$L(\sigma) = \bigcup_{r>0} B_r(\sigma). \tag{A.1}$$

Remark A.10. It is worth mentioning that the concept of local analyticity given in Definition A.8 generalizes that of Koblitz [32, p. 136], in which the local analyticity on compact $\emptyset \neq \sigma \subset \mathbb{K}$ was defined as $L(\sigma) = \bigcup_{r>0} B_r(\sigma)$.

Definition A.11 (Shnirel'man Integral). Let κ be the residue field of $\mathbb{K}$ and let $f(x)$ be a $\mathbb{K}$-valued function defined for all $x \in \mathbb{K}$ such that $\left| x - a \right| = r$ where $a \in \mathbb{K}$ and $r > 0$ with $r \in \left| \mathbb{K} \right|$. Let $\Gamma \in \mathbb{K}$ be such that $\left| \Gamma \right| = r$. Then the Shnirel'man integral of f is defined as the following limit, if it exists,

$$\int_{a, \Gamma} f(x) dx := \lim_{n \to \infty} {}' \frac{1}{n} \sum_{\eta^n = 1} f(a + \eta \Gamma), \tag{A.2}$$

where $\lim'$ indicates that the limit is taken over n such that $gcd(char(\kappa), n) = 1$.

Lemma A.12. *(i) Suppose that f is bounded on the circle $|x - a| = r$. If $\int_{a,\Gamma} f(x)dx$ exists, then $\left| \int_{a,\Gamma} f(x)dx \right| \leq \max_{|x-a|=r} |f(x)|$.*

(ii) The integral $\int_{a,\Gamma}$ commutes with limits of functions which are uniform limits on $\{x \in \mathbb{K} : |x - a| = r\}$.

(iii) If $r_1 \leq r \leq r_2$ and $f(x)$ is given by a convergent Laurent series $\sum_{k \in \mathbb{Z}} c_k(x - a)^k$

in the annulus $r_1 \leq |x - a| \leq r_2$, then

$$\int_{a,\Gamma} f(x)dx = c_0$$

and is independent of the choice of Γ with $|\Gamma| = r$, as long as $r_1 \leq r \leq r_2$. More generally,

$$\int_{a,\Gamma} \frac{f(x)}{(x-a)^k} dx = c_k.$$

Proof. The proof of statements (i) and (ii) follow directly from the definition of the Shnirel'man integral. To prove (iii), note that for $k \neq 0$ and $n > |k|$,

$$\sum_{\eta^n=1} \eta^k = 0, \tag{A.3}$$

and hence

$$f(a + \Gamma\eta) = c_0 + \sum_{k \in \mathbb{Z} \setminus \{0\}} c_k \Gamma^k \eta^k.$$

The result is now a consequence of Eq. (A.3), Eq. (A.2) and the fact that $\lim_{k \to \infty} c_k \Gamma^k = 0$.

Lemma A.13. *Fix $x_0 \in \mathbb{K}$ and $m > 0$. Then, the following holds*

$$\int_{a,\Gamma} \frac{dx}{(x - x_0)^m} = \begin{cases} 0 & \text{if } |a - x_0| < r; \\ (a - x_0)^{-m} & \text{if } |a - x_0| > r. \end{cases}$$

Corollary A.14. *Fix $x_0 \in \mathbb{K}$ and $m > 1$. Then, the following hold*

$$\int_{a,\Gamma} \frac{x-a}{(x-x_0)}\, dx = \begin{cases} 1 \text{ if } \left|a - x_0\right| < r; \\[2ex] 0 \text{ if } \left|a - x_0\right| > r. \end{cases}$$

and

$$\int_{a,\Gamma} \frac{x-a}{(x-x_0)^m}\, dx = \begin{cases} 0 \text{ if } \left|a - x_0\right| < r; \\[2ex] 0 \text{ if } \left|a - x_0\right| > r. \end{cases}$$

Lemma A.15 (Non-archimedean Cauchy integral formula [54]). *If f is analytic on $D(a, r)$ and if $\left|\Gamma\right| = r \in \left|\mathbb{K}\right|$, then*

$$\int_{a,\Gamma} \frac{f(x)(x-a)}{(x-x_0)}\, dx = \begin{cases} f(x_0) \text{ if } \left|a - x_0\right| < r, \\[2ex] 0 \quad\;\; \text{ if } \left|a - x_0\right| > r. \end{cases} \tag{A.4}$$

In particular, this integral does not depend on the choice of a, Γ or r as long as $\left|x_0 - x\right|$ is either less than or greater than r.

Theorem A.16 (Non-archimedean Residue Theorem [54]). *Let f be a rational function over $\mathbb{K}$ and suppose none of the poles x_0 off satisfy $\left|x_0 - a\right| = \left|\Gamma\right|$, where $\Gamma \in \mathbb{K} - \{0\}$. Then*

$$\int_{a,\Gamma} f(x)(x-a)\, dx = \sum_{\left|x_0 - a\right| < \left|\Gamma\right|} res_{x=x_0} f(x), \tag{A.5}$$

where $res_{x=x_0} f(x)$ is the coefficient of $(x - x_0)^{-1}$ in the Laurent expansion of f about x_0.

We next state some interesting results on the Shnirel'man integral.

A.1   Distributions with Compact Support

Let $\sigma \subset \mathbb{K}$ be a compact subset and let $r > 0$. It follows that there is a finite set I with $a_i \in \mathbb{K}$ for $i \in I$ with

$$\sigma \subset \bigsqcup_{i \in I} D(a_i, r^-), \quad \text{and} \quad D(\sigma, r^-) = \bigsqcup_{i \in I} D(a_i, r^-). \tag{A.6}$$

Let $f \in B_r(\sigma)$. Then $f : D(\sigma, r^-) \to \mathbb{K}$ is clearly analytic and hence satisfies:

(i) $f(x) = \sum_{j \in \mathbb{N}} f_{ij}(x - a_i)^j$ for $x \in D(a_i, r^-)$.

(ii) For all $i \in I$, $\left| f_{ij} \right| r^j \to 0$ as $j \to \infty$.

(iii) The norm of f is defined as $\left\| f \right\|_r = \sup_{i \in I, j \in \mathbb{N}} \left| f_{ij} \right| r^j$.

It is not hard to check that $\left(B_r(\sigma), \left\| . \right\|_r \right)$ is a non-archimedean Banach space. Moreover, the following embedding is continuous

$$B_r(\sigma) \hookrightarrow B_{r_1}(\sigma)$$

with $0 < r_1 < r$.

Recall that (from Theorem A.9) the following holds

$$L(\sigma) = \bigcup_{r > 0} B_r(\sigma).$$

Definition A.17. The space $L^*(\sigma) := L(\sigma)^*$ (topological dual of $L(\sigma)$) is called the space of distributions with support σ.

For all $\mu \in L^*(\sigma)$ and $f \in L(\sigma)$, we represent the canonical pairing between μ and f as

$$(\mu, f) = (\mu(x), f(x)) = \mu(f).$$

Moreover, it is easy to see that for $\mu \in L^*(\sigma)$, then $\mu_{|B_r(\sigma)}$ is a continuous linear functional whose norm is denoted by

$$|||\mu|||_r := \sup_{f \in B_r(\sigma), f \neq 0} \frac{\left| \mu(f) \right|}{\left\| f \right\|_r}.$$

In particular, if $0 < r_1 < r$, then

$$|||\mu|||_{r_1} \geq |||\mu|||_r.$$

For $r > 0$, $i \in I, j \in \mathbb{N}$ and $x \in \mathbb{K}$, we define

$$\chi(r,i,j;x) = \begin{cases} (x - a_i)^j & \left| x - a_i \right| < r, \\ \\ 0 & \left| x - a_i \right| \geq 0. \end{cases}$$

Obviously, $\chi(r,i,j;\cdot) \in B_r(\sigma)$.

It can also be shown (see [32]) that the weak topology on $L^*(\sigma)$ whose basis is the neighborhoods of zero given by

$$U_{f,\varepsilon} := \left\{ \mu \in L^*(\sigma) : \left| \mu(f) \right| < \varepsilon \right\}$$

and the stronger topology on $L^*(\sigma)$ whose basis is the neighborhoods of zero given by

$$U(r,\varepsilon) := \left\{ \mu \in L^*(\sigma) : \left\| \mu \right\|_r < \varepsilon \right\}.$$

have the same convergent sequences.

Throughout, we set $\overline{\sigma} := \mathbb{K} - \sigma$ and

$$\overline{D}(\sigma,r) = \mathbb{K} - D(\sigma, r^-) = \left\{ x \in \mathbb{K} : \quad dist(x,\sigma) \geq r \right\}.$$

Definition A.18. The collection of all functions $\varphi : \overline{\sigma} \to \mathbb{K}$ which are Krasner analytic and vanish at infinity that is:

(i) φ is a limit of rational functions whose poles are contained in σ, the limit being uniform in any set of the form $\overline{D}(\sigma,r)$;

(ii) $\lim\limits_{|z| \to \infty} \varphi(z) = 0$;

is denoted $H_0(\overline{\sigma})$.

For $\varphi \in H_0(\overline{\sigma})$, we define

$$\left\| \phi \right\|_r := \max_{z \in \overline{D}(\sigma,r)} \left| \phi(z) \right| = \max_{dist(z,\sigma)=r} \left| \phi(z) \right|.$$

In particular, for $0 < r_1 < r$, then

$$\left\| \phi \right\|_r \leq \left\| \phi \right\|_{r_1}.$$

As a topology on $H_0(\overline{\sigma})$, we take as a basis the open neighborhoods of zero given by

$$U_0(r,\varepsilon) = \left\{ \phi : \left\| \phi \right\|_r < \varepsilon \right\}.$$

A.2   Cauchy-Stieltjes and Vishik Transforms

Definition A.19 (Cauchy-Stieltjes Transform). Let $\sigma \subset \mathbb{K}$ be a compact subset and let $\mu \in L^*(\sigma)$. The Cauchy-Stieltjes transform of μ is the function

$$\varphi = S\mu : \overline{\sigma} \to \mathbb{K}$$

$$z \mapsto \left(\mu(x), \frac{1}{z - x} \right).$$

Let $f \in B_r(\sigma)$ and suppose $\sigma \subset \bigsqcup_{i \in I} D(a_i, r^-)$ where I is a finite index set. Fix $\Gamma \in \mathbb{K}$ such that for all $i \in I$,

$$\sup_{b \in D(a_i, r^-) \cap \sigma} |a_i - b| < |\Gamma| < r. \tag{A.7}$$

Definition A.20. We define the Vishik transform V (under the assumptions leading to Eq. (A.7)) by

$$V\varphi : B_r(\sigma) \to \mathbb{K}$$

$$f \mapsto \sum_{i \in I} \int_{a_i, \Gamma} (z - a_i)\varphi(z)f(z)dz.$$

Lemma A.21 ([54]). *Let $\mu \in L^*(\sigma)$ be a distribution with compact support. Then $S\mu \in H_0(\overline{\sigma})$ and $S : L^*(\sigma) \to H_0(\overline{\sigma})$ is continuous.*

Lemma A.22 ([54]). *Let $\varphi \in H_0(\overline{\sigma})$. Then,*

$$V\varphi(f) := \sum_{i \in I} \int_{a_i, \Gamma} (z - a_i)\varphi(z)f(z)dz, \quad f \in B_r(\sigma),$$

does not depend on the choice of a_i and Γ satisfying (A.7). Furthermore, it is compatible with the inclusion

$$B_r(\sigma) \hookrightarrow B_{r_1}(\sigma) \quad for \ r_1 < r.$$

In addition, both $V\varphi : B_r(\sigma) \to \mathbb{K}$ and $V : H_0(\overline{\sigma}) \to L^*(\sigma)$ are continuous.

Lemma A.23 ([54]). *We have $VS = SV = Id$.*

References

1. J. Araujo, C. Perez-Garcia, S. Vega, Preservation of the index of p-adic linear operators under compact perturbations. Compositio Mathematica **118**, 291–303 (1999)
2. E. Artin, *Algebraic Numbers and Algebraic Functions* (Gordon and Breach Science Publishers Inc, New York, 1967)
3. D. Attimu, T. Diagana, Functional calculus for a class of unbounded linear operators on some non-Archimedean Banach spaces. Comment. Math. Univ. Carolin. **50**(1), 37–60 (2009)
4. D. Attimu, Linear operators on some non-archimedean Hilbert spaces and their spectral theory. PhD Thesis, Howard University, Washington DC (2008)
5. D. Attimu, T. Diagana, Representation of bilinear forms in non-Archimedian Hilbert spaces. Comment. Math. Univ. Carolin. **48**(3), 431–442 (2007)
6. D. Attimu, T. Diagana, Representation of bilinear forms by linear operators in non-archimedean Hilbert space equipped with a Krull valuation. Red. Sem. Mat. Univ. Pol. Torino. **68**(2), 139–155 (2010)
7. R. Baker, A certain p-adic spectral theorem (2007). arXiv.math /070353901 [MATH.FA]
8. S. Basu, T. Diagana, F. Ramaroson, A *p*-adic version of Hilbert-Schmidt operators and applications. J. Anal. Appl. **2**(3), 173–188 (2004)
9. V.G. Berkovich, *Spectral Theory and Analytic Geometry over Non-archimedean Fields*. Mathematical Surveys and Monographs, vol. 33 (The American Mathematical Society, Providence, 1990)
10. J.W.S. Cassels, *Local Fields* (Cambridge University Press, Cambridge/New York, 1986)
11. E.B. Davies, *Spectral Theory and Differential Operators* (Cambridge University Press, Cambridge/New York, 1995)
12. T. Diagana, Towards a theory of some unbounded linear operators on *p*-adic Hilbert spaces and applications. Ann. Math. Blaise Pascal **12**(1), 205–222 (2005)
13. T. Diagana, *Non-archimedean Linear Operators and Applications* (Nova Science Publishers, Inc., Huntington/New York, 2007)
14. T. Diagana, G.D. McNeal, Spectral analysis for rank one perturbations of diagonal operators in non-archimedean Hilbert space. Comment. Math. Univ. Carolin. **50**(3), 385–400 (2009)
15. T. Diagana, G.D. McNeal, Corrigendum to "Spectral analysis for rank one perturbations of diagonal operators in non-archimedean Hilbert space". Comment. Math. Univ. Carolin. **50**(4), 637–638 (2009)
16. T. Diagana, *An Introduction to Classical and p-Adic Theory of Linear Operators and Applications* (Nova Science Publishers, New York, 2006)

© The Author(s) 2016
T. Diagana, F. Ramaroson, *Non-Archimedean Operator Theory*,
SpringerBriefs in Mathematics, DOI 10.1007/978-3-319-27323-5

17. T. Diagana, F. Ramaroson, Spectral theory for finite rank perturbations of unbounded diagonal operators in non-archimedean Hilbert space, in *Contemporary Matematics* (American Mathematical Society, To Appear)
18. T. Diagana, R. Kerby, T.H. Miabey, F. Ramaroson, Spectral analysis for finite rank perturbations of diagonal operators in non-archimedean Hilbert space. p-Adic Numbers Ultrametric Anal. Appl. **6**(3), 171—187 (2014)
19. B. Diarra, An operator on some ultrametric Hilbert Spaces. J. Anal. **6**, 55–74 (1998)
20. B. Diarra, *Geometry of the p-Adic Hilbert Spaces* (1999, Preprint)
21. B. Diarra, Bounded linear operators on ultrametric Hilbert spaces. Afr Diaspora J. Math. **8**(2), 173–181 (2009)
22. B. Diarra, S. Ludkovsky, Spectral integration and spectral theory for non-archimedean Banach spaces. Int. J. Math. Math. Sci. **31**(7), 421–442 (2002)
23. O. Endler, *Valuation Theory* (Springer, Heidelberg, 1972)
24. A. Escassut, *Ultrametric Banach Algebras* (World Scientific, Singapore, 2003)
25. Q. Fang, J. Xia, Invariant subspaces for certain finite-rank perturbations of diagonal operators. J. Funct. Anal. **263**(5), 1356–1377 (2012)
26. C. Fois, I.B. Jung, E. Ko, C. Pearcy, On rank one perturbations of normal operators. J. Funct. Anal. **253**(2), 628–646 (2007)
27. I. Gohberg, S. Goldberg, M.A. Kaashoek, *Basic Classes of Linear Operators* (Basel/Boston, Birkhäuser Verlag, 1990)
28. I. Gohberg, S. Goldberg, M.A. Kaashoek, *Classes of Linear Operators*, vol. I (Basel/Boston, Birkhäuser, 2003)
29. E. Ionascu, Rank-one perturbations of diagonal operators. Integral Equ. Oper. Theory **39**, 421–440 (2001)
30. H.A. Keller, H. Ochsenius, Bounded operators on non-archimedean orthomodular spaces. Math. Slovaca **45**(4), 413–434 (1995)
31. A.Y. Khrennikov, *p-Adic Valued Distributions in Mathematical Physics*. Mathematics and Its Applications, vol. 309 (Kluwer Academic, Dordrecht, 1994)
32. N. Koblitz, *p-adic Analysis: A Short Course on Recent Work* (Cambridge University Press, Cambridge, 1980)
33. A.N. Kochubei, On some classes of non-Archimedean operator algebras. Contemp. Math. **596**, 133–148 (2013)
34. A.N. Kochubei, Non-Archimedean unitary operators. Methods Funct. Anal. Topol. **17**(3), 219–224 (2011)
35. A.N. Kochubei, Non-Archimedean shift operators. p-Adic Numbers Ultrametric Anal. Appl. **2**(3), 260–264 (2010)
36. A.N. Kochubei, Non-Archimedean normal operators. J. Math. Phys. **51**(2), 023526, 15pp (2010)
37. M. Krasner, Prolongement analytique uniforme et multiforme dans les coprs valués complets. Colloque. Int. CNRS **143**, 97–142 (1966). Paris
38. R. Miranda, *Algebraic Curves and Riemann Surfaces*. Graduate Studies in Mathematics, vol. 5 (Providence, American Mathematical Society, 1995)
39. J. Martínez-Maurica, T. Pellon, C. Perez-Garcia, Some characterizations of *p*-adic semi-Fredholm operators. Ann. Mat. Pura Appl. **156**, 243–251 (1990)
40. G.D. McNeal, Spectral analysis for rank one perturbations of diagonal operators in non-archimedean Hilbert space, Howard University (2009)
41. H. Ochsenius, W.H. Schikhof, *Banach Spaces over Fields with an Infinite Rank Valuation, p-Adic Functional Analysis* (Poznan, 1998) (Dekker, New York, 1999), pp. 233–293
42. C. Perez-Garcia, W.H. Schikhof, *Locally Convex Spaces over Non-archimedean Valued Fields* (Cambridge University Press, Cambridge, 2010)
43. C. Perez-Garcia, S. Vega, Perturbation theory of *p*-adic Fredholm and semi-Fredholm operators. Indag. Math. (N.S.) **15**(1), 115–127 (2004)
44. C. Perez-Garcia, Semi-Fredholm operators and the Calkin algebra in *p*-adic analysis. I, II. Bull. Soc. Math. Belg. Sér. B **42**(1), 69–101 (1990)

45. P. Ribenboim, *The Theory of Classical Valuations* (Springer, New York, 1999)
46. W.H. Schikhof, *Ultrametric Calculus: An Introduction to p-Adic Analysis* (Cambridge University Press, Cambridge/New York, 1984)
47. O. Schilling, *The Theory of Valuation* (Literary Licensing, LLC, Whitefish, 2013)
48. J.P. Serre, Completely continuous endomorphisms of p-adic Banach spaces. Publ. Math. I.H.E.S. **12**, 69–85 (1962)
49. K. Shamseddine, M. Berz, Analytical properties of power series on Levi-Civita fields. Ann. Math. Blaise Pascal **12**(2), 309–329 (2005)
50. L.G. Shnirel'man, On functions in normed, algebraically closed fields. Izv. Akad. Nauk SSSR, Ser. Mat. **2**(5–6), 487–498 (1938)
51. S. Śliwa, On Fredholm operators between non-archimedean Fréchet spaces. Compositio Mathematica **139**, 113–118 (2003)
52. P. Schneider, *Nonarchimedean Functional Analysis* (Springer, Berlin/New York, 2002)
53. A.C.M. van Rooij, *Non-archimedean Functional Analysis* (Marcel Dekker Inc, New York, 1978)
54. M. Vishik, Non-archimedean spectral theory. J. Sov. Math. **30**, 2513–2554 (1985)

Index

Symbols

$C(\mathbb{Z}_p, \mathbb{Q}_p)$, 63
$H_0(\overline{\sigma})$, 100
$M_{(p)}$, 17
$U_{(p)}$, 17
$\mathbb{Q}_p$, 63, 64
$\mathbb{Z}_p$, 63
$\mathbb{C}_p$, 26
$\mathbb{F}_p$, 26
$\mathbb{Q}_p$, 25
$\mathbb{Z}/p\mathbb{Z}$, 18
$\mathbb{Z}_{(p)}$, 17
$\mathbb{Z}_{(p)}/M_{(p)}$, 18
$\overline{\mathbb{Q}_p}$, 26
$c_0(\mathbb{K})$, 46
p-adic Hilbert space, 41, 54
p-adic numbers, 25
p-adic valuation, 17

A

accumulation point, 131
additive valuation, 15
adjoint operator, 73
algebraic closure, 33
algebraic integer, 21
algebraic number field, 21
algebraically closed, 25
analytic bounded linear operator, 102
archimedean, 5

B

bicontinuously, 59
bounded linear operator, 62

C

canonical basis, 54, 63, 126
Cauchy–Schwarz inequality, 55
Cauchy–Stieltjes transform, 98
closable operator, 125
closed operator, 124
complete metric space, 8
completely continuous, 77, 78, 128, 131
continuous linear operator, 62
continuous spectrum, 82, 127

D

Dedekind domain, 21
dense, 8
dense valuation, 8
diagonal operator, 78, 82, 132
discrete valuation, 8, 27
distribution with compact support, 98

E

eigenvalue, 81, 126
eigenvector, 126
essential spectrum, 82, 107, 127, 135

F

field extension, 12
finite field, 12
finite rank operator, 75
Fredholm operator, 79, 80, 82, 107, 127, 128, 131
free Banach space, 65

© The Author(s) 2016
T. Diagana, F. Ramaroson, *Non-Archimedean Operator Theory*,
SpringerBriefs in Mathematics, DOI 10.1007/978-3-319-27323-5

G
graph norm, 128
group of units, 8

H
Hamel basis, 126

I
infinite cyclic, 8, 15
integral domain, 21
inverse operator, 67
invertible, 67

K
kernel, 67
Krasner analytic, 97
Krull valuation, 36

L
Laurent series, 19
linear operator, 61
local ring, 8

M
maximal ideal, 8, 16, 19
meromorphic function, 24
metric space, 6
multiplication operator, 63
multiplicative group, 8

N
non-archimedean, 5
non-archimedean Cauchy integral formula,
       93
non-archimedean norm, 41
non-archimedean residue theorem, 94
non-archimedean valuation, 13
non-archimedean valued field, 1
non-degenerate, 55
norm operator, 62
normal operator, 74

O
open mapping theorem, 67
order function, 15
orthogonal basis, 53, 57, 68

P
point spectrum, 81, 127
pointwise convergence, 65
prime element, 12
Puiseux series, 26

R
range, 67
residual spectrum, 82, 127
residue class field, 14, 23
residue field, 25
resolvent, 126
resolvent set, 81
Riemann surface, 24
ring of integers, 8, 16

S
self-adjoint operator, 74
series, 13
Shnirel'man integral, 91
spectral analysis, 107
spectrum, 81, 126, 133
spherically complete, 31
strong triangle inequality, 41
strongly orthogonal basis, 69, 71

T
torsion-free, 36
triangle inequality, 2, 4
trivial valuation, 2

U
ultrametric, 41
ultrametric inequality, 4
unbounded linear operator, 123
unbounded linear operators, 131
uniformizer, 12, 14, 16, 19, 25
unitary operator, 74

V
valuation, 1, 16
valuation ring, 8, 23
value group, 8
Vishik spectral theorem, 104
Vishik transform, 98
von Neumann series, 67

Printed in the United States
By Bookmasters